ENGAGING SCIENCE

ALSO BY JOSEPH ROUSE

Knowledge and Power: Toward a Political Philosophy of Science

ENGAGING SCIENCE

How to Understand Its Practices Philosophically

Joseph Rouse

CORNELL UNIVERSITY PRESS

Ithaca and London

First published 1996 by Cornell University Press.

Printed in the United States of America

Library of Congress Cataloging-in-Publication Data

Rouse, Joseph, 1952–
Engaging science : how to understand its practices philosophically / Joseph Rouse.
p. cm.
Includes bibliographical references and index.
ISBN 0-8014-3193-X (cloth : alk. paper).—ISBN 0-8014-8289-5 (pbk. : alk. paper)
1. Science—Philosophy. 2. Science—Social aspects. 3. Philosophy, Modern. I. Title.
Q175.R5665 1995
501—dc20 95-23439

⊗ The paper in this book meets the minimum requirements of the American National Standard for Information Sciences—Permanence of Paper for Printed Library Materials, ANSI Z39.48-1984.

Contents

Acknowledgments

During the seven years in which this book was conceived and written, I have been sustained by the support of colleagues, friends, institutions, and family. I begin these acknowledgments with the most direct and basic support: the National Science Foundation's Grant DIR 91-22683 provided a crucial period of released time in which Part II was substantially reconceived and partially written. I owe special thanks to Program Director Ronald Overmann for his encouragement and support throughout the process of application and review. Participation in Arthur Fine's NEH Summer Seminar, "The Legacy of Realism," and Hubert Dreyfus and David Hoy's NEH Summer Institute, "Interpretation and the Human Sciences," provided initial stimulation. Discussions in the NEH Summer Institute "Science as Culture and Practice," which I codirected with Steve Fuller at Wesleyan University, both challenged and helped consolidate my thinking about the project. Sabbaticals from Wesleyan University in 1989 and 1991 and a Faculty Fellowship in 1988 from Wesleyan's Center for the Humanities provided invaluable opportunities for reflection and writing.

In this book, I insist on the importance of the interactive and dynamic aspects of knowledge. Fittingly, therefore, I note with gratitude that much of what I have written makes best sense as contributions to continuing conversations with Arthur Fine, Steve Fuller, Donna Haraway, Mark Okrent, and Betty Smocovitis. They must be absolved of blame for anything written here, but they are nevertheless causally responsible for much of it. I thank many others for contributions that exceed what can be specifically acknowledged in footnotes; among them are James Bohman, Judith Butler, David Carr, Nancy Cartwright, John Compton, Robert Crease, Hubert Dreyfus, Brian Fay, Nancy Fraser, Peter Galison, Lydia Goehr, Gary Gutting, Sandra Harding, David Hiley, Andrew Pickering, Hans-Jörg Rheinberger, Paul Roth, Thomas Ryckman, David Stump, Kenneth Taylor, Steven Vogel, Thomas Wartenburg, Kenneth Westphal, Samuel Wheeler III, Steve Woolgar, and Alison Wylie. The book has been significantly improved as a result of comments on earlier versions of some chapters from Margaret Crouch, Brian Fay, Ruth Ginzberg, Lydia Goehr, Steven Horst, Andrew Norman, Mark Okrent, Thomas Ryckman, Mark Stone, and two anonymous reviewers for Cornell University Press. Roger Haydon has been a judicious and helpful editor.

As always, my students at Wesleyan University helped me work through much of what I say here, and their always challenging questions, objections, clarifications, and constructive responses have been indispensable. Special thanks go to Max Belkin, Ed Boedeker, Dave Burland, Josh Dunsby, Laura Fedolfi, Henry Goldschmidt, Jo-Jo Koo, Jennifer Langdon, Elizabeth Merchant, Susannah Paletz, Jessica Pfeifer, Josh Schwimmer, Margaret Sewall, Peter Spillman, and Chris Zurn.

The book has also benefited greatly from opportunities to present portions to thoughtful and responsive audiences, including the philosophy departments at the University of Connecticut, Denison College, the University of Georgia, the University of New Hampshire, Oberlin College, Saint Louis University, the State University of New York at Stony Brook, the University of Western Ontario, the College of William and Mary, and also the American Philosophical Association's Eastern Division, the Pembroke Center at Brown University, the Science Studies Program at the University of California at San Diego, the Program in History and Philosophy of Science at the University of Colorado, the History Department at the University of Florida, the "Disunity and Contextualism" conference at Stanford University, the Maxwell Graduate Center at Syracuse University, the Center for the Humanities at Wesleyan University, the Minnesota Center for the Philosophy of Science, and the Van Leer Institute of Jerusalem in conjunction with the Hebrew University and the University of Tel Aviv. I am especially grateful for invitations to speak at the UCLA conference "Located Knowledges" and at the Center for the

Critical Analysis of Contemporary Culture at Rutgers University, for which events the original versions of Chapters 4 and 9 were written.

I gratefully acknowledge permission to incorporate previously published material into the following chapters:

Chapter 1, reprinted from "Philosophy of Science and the Persistent Narratives of Modernity," *Studies in History and Philosophy of Science* 22 (1991): 141–62, with kind permission from Elsevier Science Ltd, The Boulevard, Langford Lane, Kidlington OX5 1GB;

Chapter 2, reprinted from *Philosophy of Science* 58 (1991): 607–27, with the permission of the Philosophy of Science Association;

Chapter 3, reprinted from *PSA 1988,* volume 1, edited by Arthur Fine and Jarrett Leplin, 294–301, with the permission of the Philosophy of Science Association;

Chapter 6, revised and reprinted from "The Narrative Reconstruction of Science," *Inquiry* 33 (1990): 179–196, by permission of Scandinavian University Press, Oslo, Norway;

Portions of Chapter 7, revised and reprinted from "The Dynamics of Power and Knowledge in Science," *Journal of Philosophy* 88 (1991): 658–67;

Chapter 9, expanded and reprinted from "What Are Cultural Studies of Scientific Knowledge?" *Configurations* 1, no. 1 (1993): 1–22. Copyright 1993 by The Johns Hopkins University Press and Society for Literature and Science.

Special thanks go to the talented and committed teachers at the Neighborhood Preschool, who provided a different kind of support and security, which has nevertheless been crucial to my work. My sons, Brian Grucan Rouse and Martin Grucan Rouse, have given me invaluable perspective on matters far more important than the book. My mother, Helen Rouse, and memories of my father, J. T. Rouse, continue to inspire me. Sally Grucan has made the most important contribution of all, providing the love, encouragement, sharing, and joy that have been the setting for my life. To Sally I dedicate this book in love and admiration.

JOSEPH ROUSE

Middletown, Connecticut

ENGAGING SCIENCE

Introduction

For much of the twentieth century, philosophical discussion of science has been shaped by the successive formation of at least four broad interpretive traditions for making sense of scientific knowledge. The disciplinary origins of philosophy of science, at least in the Anglo-American world, can be traced to the Vienna Circle, and the forced prewar emigrations that established the logical empiricist and Popperian programs within the United States and Great Britain. Subsequently, historical rationalism, scientific realism, and social constructivism have presented themselves as serious alternatives to the widely perceived failures of the positivist/empiricist program. Each of these traditions has its own internal history, marked by controversy and the gradual reshaping of central issues and doctrines. None displays a monolithic consensus. Yet for all their complexity and internal differentiation, these traditions have important common themes and fundamental shared *issues* that mark the disputes among them.[1]

1. I take the postpositivist tradition to have been inaugurated by the influential works of N. R. Hanson, Thomas Kuhn, Stephen Toulmin, Paul Feyerabend, and Michael Polanyi. Its subsequent development runs through the work of Imre Lakatos, Dudley Shapere, Mary Hesse, and Larry Laudan, among others. Scientific realism in its contemporary versions can be traced back to the work of Wilfred Sellars, J. J. C. Smart, and Grover Maxwell, but its prominence as an alternative interpretive tradition arises with the work of Hilary Putnam and Richard Boyd. It has since proliferated into a variety of moderate, internal, piecemeal, and entity realisms. The social constructivist tradition begins with the Edinburgh Strong Programme but has been substantially modified by the ensuing development of Bath relativism, ethnographic studies, discourse analysis, actor/network theory, and constitutive reflexivity. Despite the widely acknowledged death of logical positivism, there have been influential

This book responds to my serious dissatisfaction not only with these interpretive traditions individually but also with common elements in their conceptions of the issues dividing them and of what is at stake in the resolution of those issues. The chapters in Part I articulate some of the underlying assumptions and issues that motivate my dissatisfaction with these influential developments in philosophy and sociology of science. In Part II I begin to work out an alternative approach to engaging the sciences philosophically. Inevitably, however, these two sections overlap. My specific criticisms of the philosophical and sociological traditions indicate important themes to be explored in subsequent chapters, which in turn extend and draw on the earlier criticisms of the tradition. The remainder of this Introduction has four sections. In the first I offer an overview of my interpretation of the dominant interpretive traditions in philosophy and sociology of science and their shared commitment to the neediness of modern science for wholesale legitimation. In the second section I summarize my criticisms of common themes underlying the explicit differences among these interpretive traditions. The third section is a survey of some of the philosophical issues raised by reconceiving science studies on the model of interdisciplinary cultural studies. At the end of the Introduction I briefly summarize the individual chapters to indicate how each fits into this overall picture of the book.

REALISM, RATIONALITY, SOCIAL CONSTRUCTIVISM, AND THE LEGITIMATION OF MODERN SCIENCE

A central concern of both the philosophy and the sociology of science has been to make sense of the various activities that constitute scientific inquiry. This concern underlies both Hilary Putnam's claim that "the typical realist argument against idealism is that it makes the success of science a *miracle*,"[2] and Bas van Fraassen's response that "constructive empiricism makes better sense of science, and of scientific activity, than realism does, and does so without inflationary metaphysics."[3] Steven Shapin and Simon Schaffer propose yet a third

attempts to reinvigorate the positivist tradition, for example, by the quite distinct approaches of Bas van Fraassen and Clark Glymour. The work of Arthur Fine, Nancy Cartwright, and Ian Hacking also has different but important affinities to the positivist tradition, as I discuss in Chapter 4. Michael Friedman, Thomas Ryckman, Thomas Ricketts, and Alan Richardson, among others, have also shown that Vienna Circle positivism included a more interesting array of positions and arguments than the subsequent development of the logical empiricism of the 1950s might suggest.

2. Putnam 1978: 18.

3. Van Fraassen 1980: 73.

strategy for making sense of science: "As we come to recognize the conventional and artifactual status of our forms of knowing, we put ourselves in a position to realize that it is ourselves and not reality that is responsible for what we know."[4] Scientific activities make sense, it would seem, when they can be understood as rationally related to the achievement of a characteristic end. Scientific realists, for example, try to make sense of scientific practices as aiming to generate theories that are at least approximately true of a mind-independent world. Instrumentalists counter that scientific work makes better sense if we understand its goals to be less ambitious: better predictions of observable events, more extensive and precise problem solutions, or successful manipulation of things in pursuit of other ends. More historically inclined philosophers have suggested that the goal of scientific practice is self-directed: the sciences aim to improve their theories, methods, and goals over time, such that scientists learn better how to learn. Whatever end or goal is posited to make sense of science must, of course, itself be a desirable end within the contexts where scientific inquiry takes place. But most philosophers of science have taken the ends they attribute to scientific practice to be generally desirable, such that specific features of the contexts where science is practiced are of little philosophical interest. Any culture that is capable of scientific inquiry would presumably find it valuable to have (approximately) true theories, reliable predictions, or improved methods of inquiry.

On reflection, however, we are not entirely clear on what is at stake among competing philosophical interpretations of scientific knowledge. What difference would it make if we were to discover that science really is best understood instrumentally or realistically or in terms of its historically progressive rationality? The improvement in cultural self-understanding might be valuable in its own right, especially because the natural sciences are so culturally prominent. Yet an important reason to value accurate self-understanding is to guide future activities. In that respect it is striking how limited are the ambitions of so many contemporary philosophical interpreters of the sciences. Philosophy and sociology of science often invert Marx's 11th Thesis on Feuerbach: heretofore, perhaps, philosophers have attempted to change science, but in recent decades, the point has been just to interpret it.

Philosophy of science has not always been so restrained. Midway through the twentieth century, the most prominent philosophical frameworks for making sense of the sciences all adopted a critical

4. Shapin and Schaffer 1985: 344.

stance toward at least some scientific practices. Logical positivism and Popperian falsificationism allied themselves more closely with the scientific enterprise than did phenomenology or neo-Marxist critical theory, but they were nevertheless quite prepared to challenge those scientific beliefs and practices which did not satisfy their methodological standards. Indeed, their allegiance was primarily to a philosophical ideal of science, which actual scientific practices often failed to exemplify. By contrast, contemporary scientific realists may differ from an empiricist such as Bas van Fraassen or historical rationalists like Larry Laudan or Dudley Shapere in their interpretation of scientific achievements and goals, but they, too, would leave most scientific practices and their results unchanged.[5] The issue at stake among these philosophers turns out to be what epistemic attitude ought to be taken toward the achievements of scientific practice and especially toward scientific theories. Should we believe that well-established scientific theories are (likely to be) true? Should we eschew belief and accept those same theories as empirically adequate at best? Or should we endorse the pursuit of those theories without committing ourselves to attitudes of acceptance or belief? Scientists themselves might quite understandably be indifferent toward these interpretive disputes or perhaps choose among the competing positions aesthetically since these positions place no constraints on and offer no advice for scientific practice.

But perhaps we fundamentally misunderstand the significance of these philosophical disagreements if we look at them as merely intramural disputes among philosophers who want to clarify a shared commitment to the worth of the scientific enterprise. The contemporary situation within the philosophy of science has been decisively shaped by its encounter with the work of Thomas Kuhn and Paul Feyerabend. Whatever Kuhn and Feyerabend themselves would say about the matter, their work has been widely interpreted within philosophy and elsewhere as an attack on the rationality and cultural authority of the sciences.[6] Moreover, this positioning of Kuhn and

5. Perhaps surprisingly, this quietism is also characteristic of much of the sociological literature, which purports to substitute a purely *descriptive* approach to interpreting scientific knowledge for philosophers' *defense* of contemporary scientific practice. I discuss this feature of the social constructivist literature later in this introduction.

6. It is now widely recognized that this image of Kuhn as an "irrationalist" is quite at odds with Kuhn's own self-understanding and that the motivations and goals of Feyerabend's (1975) criticisms of the alleged rationality of science have also frequently been misconstrued. For their own positive responses to such misinterpretations, see Kuhn 1977 and Feyerabend 1987: 19–89. Responses to these rectifications nevertheless have often been typified by Alan Musgrave's remark about Kuhn: "Perhaps the revolutionary never really existed—but then it was necessary to invent him" (Musgrave 1980: 51).

Feyerabend has been enthusiastically taken up within much of the sociology of science within the past two decades. Beginning with the Edinburgh Strong Programme and the Bath constructivist-relativist program, many sociologists of science have suggested that the same phenomena that philosophers explain in terms of the rationality of scientific practice or the approximate truth of its results can be better accounted for socially.

This contrast between philosophical and sociological approaches to science initially seems quite significant, because the sociologists' proposed explanations would not legitimate the authority frequently conferred on scientific practices and results. It is one thing to argue that scientific results are approximately true, empirically reliable, or the rational outcome of inquiry; if borne out, such arguments offer reasons to accept those results and to support the practices and methods that helped achieve them. But if those results are better explained by the interests of prevailing social groups or by other social contingencies of the context in which they were established, then at best the explanation is neutral with respect to whether those results ought to be accepted. To the extent that scientific results have been widely accepted *because* they were thought to be approximately true, or rationally justified, the success of such sociological explanations might go further and actually undercut their justification. The stakes at issue in philosophical interpretations of science may therefore acquire a quite different significance in the contrasts to sociological accounts of scientific knowledge and to the legacy of Kuhn and Feyerabend. It would be important to make the best (rational) sense of scientific inquiry in order thereby to defend the authority and perhaps the cultural and political autonomy of science. A less adequate interpretation would be more vulnerable to criticisms of the rationality and authority of scientific claims to knowledge.

On first consideration, then, the principal issues between the proponents of scientific realism or the rationality of scientific methods and their sociological critics seem relatively clear. Is science a distinctively rational form of inquiry, one to which any sensible person owes some deference in attempting to understand and cope with the world? Or is scientific practice a social enterprise conducted according to local norms, involving conflicting interests and interpretations, domination, resistance, negotiation, and compromise, which in the end is no more (but also presumably no less) rational or reliable than any other complex social practice?

The stakes in these conflicts also seem initially clear and compelling. Philosophical defenders of the authority and autonomy of the sciences frequently allude to the most extreme dangers of epistemological relativism. If scientific results were neither approximately true nor the outcome of characteristically rational methods of inquiry and criticism, then no good grounds exist on which to prefer one set of beliefs about the world to another. Evolutionary biology would have no legitimate claim to supplant creation science within public school curricula; quackery and medical science would have equal epistemic standing; there would be no rational grounds for criticizing "scientific" racism or sexism; and the many apparently successful technological applications of science would be inexplicable and miraculous. A defense of the rationality or approximate truth of scientific theories and methods is thus put forward as indispensable both to the legitimate practice of scientific inquiry and to an adequately reflective and self-critical cultural politics.

The new sociologists of science in turn discount the supposed dangers of relativism and insist instead on the importance of criticizing the cultural hegemony of the sciences. The epistemic or methodological relativism characteristic of the new sociology of science has been distinguished from a "judgmental relativism" that would take all knowledge claims to be equally valid.[7] To say that the justification of scientific standards, values, and norms is relative to a particular community or social form of life supposedly does not by itself challenge their application within their appropriate context. For the purposes of everyday practice, justification relative to a specific social context is justification enough. Yet within the larger cultural contexts where it is important to engage in critical reflection on scientific knowledge, methodological relativism, or a principle of explanatory symmetry between beliefs that are accepted within scientific communities and those which are not, may seem essential. The sciences have enormous prestige and authority in the culture of the West. To understand how science is practiced and how its results are achieved and recognized, the sociologists insist, we must bracket any presumption of the rationality of scientific knowledge, allowing its achievements to present themselves on a par with other beliefs and practices. Only then might the deeply entrenched cultural presumptions in favor of scientific authority be overcome sufficiently to enable clearheaded judgment about the epistemological standing of scientific practices and beliefs. Only then, in turn, could one adequately assess the place of scientific inquiries and achievements within the many cultural settings in which they have been influential.

7. Knorr-Cetina and Mulkay 1983: 5–6.

The conflicts between philosophical and sociological interpretations of scientific practice thus reveal conflicting impressions of the situation to which those interpretations must respond. Philosophical interpretations of science acquire their cultural significance from a felt need to defend the rationality of science and its epistemic achievements against a widespread disregard for the value of reasoning and evidence and an unreasoned hostility to science. Sociological relativism responds to an alternative picture, in which general appeals to the rationality of scientific practices or the approximate truth of its theories serve the ideological function of concealing the unjustified cultural hegemony of the sciences. Sociological accounts aim to show how scientific authority is constructed contingently and thereby to undermine the widespread illusion of natural or rational necessity which has often been associated with scientific theories and concepts.

These are familiar images of our cultural situation, but their very familiarity also invites suspicion. In Part I, I discuss parallels between these debates over the legitimation of the sciences and the cultural "metanarratives" that defend or reject "the legitimacy of the modern age." The narrative structures within which the principal philosophical and sociological interpretations of science make sense fit quite comfortably within this latter genre. Philosophers have reconstructed the history of the sciences as a story of the maturation of rational scientific methods and the growth of scientific knowledge. Such progress stories are an integral part of the celebration of modernization as the progressive triumph of human reason over the irrational forces of tradition, reaction, and superstition. Sociological interpretations of science have taken a more ambivalent stance toward such celebrations. As I note in Chapter 1, the most influential sociological programs of the last few decades have sometimes aligned themselves with antimodernist criticisms of rationality and progress. More frequently, however, they have taken on a hypermodernist stance: philosophers' and scientists' faith in the distinctive rationality and progressiveness of the sciences is yet another irrational dogma that must finally succumb to (sociological) reason. What is striking is that either way, the sociologists share with the philosophers a sense of the sciences as integral to the historical development of a characteristically "modern" world, a development that fundamentally needs criticism, defense, or completion.

But why be suspicious of these grand cultural narratives? What is wrong with situating the interpretation of scientific knowledge squarely in the midst of fundamental questions about our present cultural predicament? Leave aside for the moment the adequacy of the larger narrative frameworks that constitute our world as "modern," even though very serious questions arise there. Ask instead how these stories of modernity frame the interpretation of science and scientific knowledge. For it is striking how the apparently conflicting interpretations of science offered by the positivist, historical rationalist, realist, and constructivist traditions share some fundamental assumptions about science, the interpretation of science, and what is at stake in their conflicting interpretations. These shared assumptions include the scale or generality, the historiographic locus, and the abstractness of the legitimation question that confronts the sciences. More fundamentally, they include a shared focus on scientific knowledge as what is at issue in the interpretation and legitimation of science, a shared conception of knowledge as representational, and a shared sense of what is at issue in explaining or accounting for knowledge.

Begin with the general underlying agreement among the various interpretations that the question of legitimation arises for science as a whole. For the logical positivists and for Popper, legitimation was necessarily wholesale, because the legitimating features of scientific knowledge were to provide a principled demarcation of the boundaries of science and of empirical knowledge more generally. Their successors have downplayed the demarcation project; nevertheless, they still frame global arguments to address questions of legitimation. Scientific realists, for example, argue that in a "mature" science, the best explanation for the instrumental success of its theory-dependent methods is the approximate truth of its theories. Thus, correspondence to "a reality largely independent of our thoughts and theoretical commitments" is a characteristic achievement of the most recognized scientific fields.[8] Van Fraassen offers similarly general reasons for construing *the* aim of all scientific inquiry to be empirically adequate theories. In both cases, a general interpretation is proposed to make sense of the sciences as successful practices and thereby to account for (in the dual sense of explaining and justifying) their cultural authority and autonomy.

The historical rationalist tradition may seem to move away from such global interpretations of science, for its adherents increasingly advocate making sense of science "piecemeal." But as I argue in Chapter 1, in the supposedly domain-specific interpretations of science they propose, Shapere, Laudan, Richard Miller, or even Peter

8. Boyd 1984: 42; it is also Boyd who characteristically uses the term 'mature science' to mark the intended scope of his arguments.

Galison still develop general historiographical frameworks that situate the history of any particular scientific discipline or domain in order to assess the legitimacy of its achievements.[9] The telltale marks of the "worship of generality" and the felt need to speak to the legitimacy of the modern age are manifest throughout the historiography of the "piecemeal" programs.

Social constructivists frequently demand an equally wholesale rejection of the philosophical legitimation of the scientific enterprise. The two loci classici for constructivism as a historiographical program are David Bloor's four tenets of the "Strong Programme" and Steven Shapin's riposte to Imre Lakatos's historiographical claims.[10] Bloor insists on a general principle of symmetry which would require the same kinds of causal explanations to account for apparently true and false beliefs, and he suggests that only sociological explanations could plausibly satisfy this demand. There has been considerable variation among sociologists' subsequent accounts of what constraints symmetry imposes on the interpretation of science. For example, Harry Collins extended Bloor's principle beyond its original intent when he "argued that one implication of symmetry is that the natural world must be treated as though it did not affect our perception of it."[11] Others have challenged the explanatory emphasis on macrosocial interests which characterized the work of Bloor and the Edinburgh school, preferring to account for knowledge claims on the basis of microsocial interactions.[12] But I take the core of a "social constructivist" account to be twofold: first, all scientific beliefs must be accounted for by *social* factors, whatever that analytical category turns out to include; second, any adequate interpretation of scientific knowledge claims must be neutral with respect to their epistemic or political legitimacy and hence to that extent is committed to some form of epistemic relativism.[13]

9. Shapere 1984, Laudan 1984, Miller 1987, Galison 1988. The phrase "worship of generality" in the next sentence is taken from Miller. Fine (1991) has also argued for the historiographical generality of Miller's and Shapere's programs.

10. Bloor 1991 and Shapin 1982.

11. Collins 1983: 88.

12. Among the classic examples of a microsociological approach to making sense of scientific knowledge are Latour and Woolgar 1986 (the word 'social' was removed from the title in the 2nd edition); Knorr-Cetina 1981; and Collins 1983, 1992.

13. This description of social constructivism is both stipulative and polemical. Although it fairly describes many of the classical positions in the sociological tradition, some influential participants in that tradition now reject one or both of these tenets (see, for example, Latour 1987; Pickering 1992, 1995). I identify "social constructivism" with what these sociologists sometimes call traditional or classical sociology of scientific knowledge ("SSK") precisely in

Sociological relativism by itself might opt out of the global legitimation or critique of scientific knowledge were it not for a third commitment held in common with many philosophical interpretations of scientific knowledge. What most fundamentally unites all the standard interpretive traditions is that the place of scientific knowledge in our culture is in *need* of global interpretive legitimation. I shall call this commitment to the neediness of science the *legitimation project.* This often tacit commitment accounts for the intended significance of philosophical and sociological interpretations of science alike. In making sense of science in a way that meets this need, the philosophers take themselves to play an important role within the larger culture of science. In proposing to account fully for scientific knowledge in ways that offer no comforting legitimation of the rationality or truth of scientific beliefs, the sociologists suggest that this need must remain unsatisfied and hence that the present cultural role of scientific knowledge is deeply suspect.

These conflicts within the legitimation project occur at a surprisingly high level of abstraction from scientific practice. When philosophers provide a general historiographical defense of the rationality of science, for example, they do not propose that the actual institutions and practices in which scientific work has taken place are especially rational or progressive. No one claims in these contexts, for example, that the sources, priorities, and disbursement of funding for scientific research have been established on a rational basis. Nor has there been any serious attempt to defend or criticize the judgment criteria, formats, or distribution patterns of scientific literature; the educational and career paths of those who train as scientists; the role of national armed forces, institutionalized medicine, the chemical or petroleum industries, or agribusiness in shaping scientific priorities; or the political activities of scientists and their influence on public policy. Yet this abstraction of the legitimation of science from its particular institutional locations seems to reflect several alternative tacit justifications, as Steve Fuller has noted:

> On the one hand, internalism can be defended (a la Lakatos) by arguing that a preferred model of scientific rationality can be instantiated by an indefinite number of sociologies, given that a methodology like falsificationism does not imply a specific way of

---

order to highlight their departure from and improvement on this sociological tradition, to make common cause with it, and to display its far-reaching implications.

> organizing knowledge production. . . . On the other hand, internalism can be defended (a la Shapere) by arguing that the preferred model of scientific rationality is simply the one that has in fact evolved during the history of science. The philosopher pitches her description so abstractly because she wants to concentrate exclusively on those features of the actual situation that have contributed to the growth of knowledge.[14]

Fuller omits a third possibility of agnosticism between these alternatives: scientific realists provide a general argument to show that the actual practices of the "mature" sciences have contingently succeeded in formulating approximately true theories, with no commitment to whether the achievement could be duplicated with a significantly different institutional history.

The social constructivists have done rather better in bringing the social practices and institutions of science into their interpretations in their many detailed case studies. In these particular cases, they discuss the considerations I mentioned above and more. Yet their contribution to the legitimation project is oddly abstracted from the details of their specific case studies. Donald MacKenzie and Barry Barnes have aptly noted this point in assessing the larger implications of their study of the controversies between Mendelians and biometricians at the turn of the century: "The general point is not that the goal-directed character of scientific judgment implies its relationship to any particular contingency, or to external factors, or political interests; what is implied is that any such contingency *may* have a bearing on judgment and that contingent sociological factors of some kind *must* have."[15] The fact of social contingency by itself, not the disreputable or dissuasive character of specific contingencies shaping the sciences, has led social constructivists to challenge the cultural authority of the sciences. When Andrew Pickering concluded his rich historical study of high-energy physics by saying that "there is no obligation upon anyone framing a view of the world to take account of what twentieth-century science has to say,"[16] his rationale was to be found not in the particular details of his sociological explanation of physicists' commitment to gauge field theories and quarks but in the availability of contingent social details for this purpose at all.

Such abstractions from particular case studies have also been sup-

14. Fuller 1989: 11.
15. MacKenzie and Barnes 1979: 205.
16. Pickering 1984: 413.

plemented by quite general arguments. Thus, Collins challenges the legitimating role often played by the replicability of scientific experiments by what he calls the "experimenter's regress": there is no measure of whether an experiment is competently performed apart from its achievement of the "correct" outcome, which must therefore be determined on other than experimental grounds. Collins then insists that the outcome of experiments *must* be explained socially as a decision undetermined by any other matter of fact.[17] Similarly, Bruno Latour and Steve Woolgar offer a general argument about the relationship between scientific statements and their purported objects, so as to deny objects any role in explaining the acceptance of statements: "The distinction between reality and local circumstances *exists only after* the statement has stabilised as a fact. . . . "Reality" cannot be used to explain why a statement becomes a fact, since it is only after it has become a fact that the effect of reality is obtained."[18] Only the scientists' "microprocessing" activities determine when a statement splits itself into a statement and an independent object to which it refers. Scientific statements themselves thus turn out to be fictions, with the most important difference between alternative fictional accounts being the resources that can be marshaled to support them.[19]

The level of generality and abstraction at which most philosophical and sociological interpreters of science have engaged in the legitimation project is directly and fundamentally connected to the common focus of various interpretive traditions on interpreting science as a knowledge-producing activity, with scientific knowledge understood to consist in some form of representation. That philosophical interpretations of science would focus primarily on scientific knowledge is not initially obvious. The philosophy of science as a subdiscipline has actually had an ambivalent relationship to the longer history of epistemology. Theories of knowledge traditionally ask, What is knowledge in general? What are the necessary and sufficient conditions for its attainment? Only then can they consider which inquiries actually satisfy these conditions. A philosophy of science begins instead by considering actual practices of inquiry. Philosophers of science then ask what these practices achieve, how they achieve it, and why those achievements are valued. Its normative dimension arises not from a prior conception of what knowledge *must* be but from a reflection on what accounts for the successes of the best practices of

17. Collins 1992.
18. Latour and Woolgar 1986: 180.
19. Latour and Woolgar 1986: 252–58.

inquiry, which can then be a benchmark for the critical assessment of others. It is not accidental that the disciplinary origins of philosophy of science can be traced to a group—the Vienna Circle—that was composed mostly of scientists, mathematicians, social scientists, and historians, that is, people who were not philosophers by training.

Yet despite its disciplinary ambivalence toward traditional epistemology, the actual practice of the philosophy of science has mostly been epistemologically conservative, as its central questions have been structured by traditional conceptions of knowledge. Michael Williams (following Barry Stroud) has offered a useful characterization of the core of the epistemological tradition: "The traditional philosophical examination of knowledge . . . points to four central ideas: an *assessment* of the *totality* of our knowledge of the world, issuing in a judgment delivered from a distinctively *detached* standpoint, and amounting to a verdict on our claim to have knowledge of an *objective* world."[20] Philosophy and sociology of science have shifted the focus of their examination from knowledge in general to scientific knowledge but have otherwise adapted these four themes to their own purposes. The legitimation project is just a commitment to the need for a detached assessment of the totality of science as claiming knowledge of an objective world.[21] Hence, we can recognize the legitimation project as an expression of the continuity between the philosophy of science and epistemology and between the sociology of science and the sociology of knowledge.

Philosophers of science have also borrowed from the modern epistemological tradition a conception of scientific knowledge as consisting of *representations* in a linguistic *medium*.[22] What needs to be

20. Williams 1991: 22.

21. There are, of course, important differences in how each of these themes is interpreted. Philosophers typically characterize "the totality of science" normatively (only "mature" sciences are included; the assessment of other, putative sciences must wait until they attain suitable maturity), whereas sociologists insist that we must treat in the same way all practices that aim for or claim authoritative knowledge. Similarly, philosophers and sociologists offer different models of an appropriately "detached" stance. Philosophical reflection proposes detachment from all practical interests and involvements except an interest in objective knowledge. Hence, philosophical detachment aims to explore an interest *shared* with the sciences it studies, but to do so in abstraction from the many other interests that animate scientific practice. Sociologists offer a different picture of detachment. Any suggestion of shared interests or norms between sociologists and the scientists they study must be bracketed or abolished in order to treat scientific practices and beliefs as objects of sociological research, just like any other social phenomena. If necessary, sociologists must actively seek estrangement from the sciences they study so as to prevent familiarity from overriding their own scientific detachment.

22. The conception of knowledge as representation is a central contribution of seventeenth-century epistemology, most notably in the work of Descartes and Locke. The linguistic turn that reformulated questions about knowledge as questions about relations among the

made sense of philosophically, then, is the three-term relation between a knower (a rational subject for whom representations are meaningful), knowledge (the content of meaningful representations and their warrant), and the objects known (what is purportedly represented by scientific knowledge). In turn, two main issues have arisen in understanding this relationship: accounting for the "content" of knowledge (including the internal structure of that content), and determining the "ontological status" of the objects of knowledge as represented by that content. Both these concepts need explication.

The notion of the "content" of scientific knowledge can be usefully connected to Gottlob Frege's conception of the sense of a linguistic expression. Frege identified the sense of a sign with its "mode of presentation" of an object, a mode that both determines which object the sign refers to and presents it in only one of many possible ways (thus, "author of *The Origin of Species*" and "naturalist on the *Beagle*" both refer to Darwin but present him differently). The sense or meaning conveyed by a sign is thus distinct both from the actual sign used (the same sign can convey different senses, and different signs can convey the same sense) and from the object indicated by the sign (the same object can be indicated in different ways, and signs can have a sense in the absence of any object that satisfies it).[23] The representational content of scientific knowledge would therefore be the sense or meaning conveyed by the various ways in which that knowledge is expressed by scientists, which in turn can be reconstructed in still different terms by historians and philosophers. In focusing their interpretations primarily on the content of knowledge, philosophers of science have hoped to extract *what* scientists have said or believed for what reasons from considerations of when, where, by whom, and to whom it was said and for what purposes.

The central questions of the philosophy of science have been addressed mainly to the inferential structure of the epistemic content of the sciences. The classical positivist questions about confirmation, explanation, reduction, and the interpretation of theoretical terms were explicitly formulated as questions about the inferential relations

semantic contents of linguistic expressions begins with Gottlob Frege and Bertrand Russell at the turn of the twentieth century. My point in emphasizing this lineage is to suggest that the mainstream tradition in the philosophy of science can be understood as the application of the central concepts of classical epistemology to scientific knowledge. The radical potential of philosophy of science—to reconceive knowledge on the basis of reflection on science—has not actually been fulfilled.

23. Frege 1970: 56–78, esp. 57–61.

among the contents of the expressions of scientific theories, laws, and observation reports which constitute the content of knowledge. Subsequent attention to scientists' methodological and evaluative commitments have also been formulated in terms of the content of their methods, values, and standards. For although many philosophers have increasingly moved away from trying to reconstruct the content of scientific knowledge as a formal system, the distinctively philosophical character of postpositivist attention to the history of science has similarly been the reconstruction of that history as a story of the dialectical relations within the evolving content of scientific knowledge.

Both scientific realists and social constructivists appear to reject the positivists' and early postpositivists' emphasis on the content of knowledge abstracted from its material and social circumstances. After all, the contemporary versions of scientific realism have been formulated specifically in opposition to Fregean accounts of meaning and reference and to the associated difficulties about meaning change which initially preoccupied postpositivist philosophy of science. Social constructivists, meanwhile, directly oppose the internalist approach to understanding science, which seems obviously connected to the philosophical focus on epistemic content. But in both cases, these are differences over how to understand and make sense of the content of scientific knowledge and not a turn away from or a criticism of the concept of "content" itself.

Scientific realists reject classical Fregean accounts of how the content of knowledge is determined and how that content is intentionally directed toward objects in the world, but the concept of epistemic content is still important for their interpretation of scientific knowledge. Realists typically deny that conventional meanings of linguistic signs establish their sense, which in turn fixes their intended reference. Instead, they argue that the reference of an expression is fixed first by the causal relations that occasioned its initial or canonical use. Thus, when J. J. Thomson used the word 'electron' in characterizing his experiments with cathode ray tubes, the word thereby referred to whatever (kind of) objects were actually causally connected to the phenomena he was discussing. The meaning content of his utterances about electrons is also determined to a significant extent by the objects themselves: their content is fixed by whatever interpretation maximizes the extent to which they truthfully represent the electrons.

Three points are worth noting about a realist approach to understanding scientific language. First, a determinate content to scientific utterances is crucial to realism about science, because the relation of correspondence (or noncorrespondence) between the content of scientists' claims and the way things are plays an explanatory role in their overall interpretation. There must be not only a fact of the matter about the way the world is but also a fact of the matter about what scientists' utterances *say* about the way the world is. Second, although realists appeal (at least in principle) to contextual features of the initial or canonical use of an expression to determine to which objects it refers, these features are then ignored thereafter. The initial baptism of the object serves to fix the reference of its denoting terms once and for all. Third, this account at bottom still presupposes something like a Fregean understanding of the relation between sense and reference for the semantically crucial concept of 'cause'. The (already fixed) sense of this word must refer directly to real, mind-independent connections between scientists' use of words and the things around them. Otherwise, the reference of 'cause' could itself never be fixed: it would refer to whatever relations (really) caused its original or canonical use, but these relations could not be determined without our already knowing to what 'cause' referred.[24]

Social constructivists may seem to offer a more radical criticism of the notion of the content of knowledge abstracted from its occasioned expressions. They typically insist on the ineliminable importance of local contextual features of knowledge production for making sense of scientific work. Yet, perhaps because it has been so important for them to distinguish their approach from earlier functionalist sociological interpretations of science, constructivists have repeatedly emphasized the importance of including the content of scientific knowledge within the domain of sociological explanation. Even Latour, who also criticizes many of the same aspects of social constructivism as I do, in a 1990 review repeatedly chastised Sharon Traweek's work for being "unable to relate the content of physics to the social organization."[25] As Fuller concluded:

> The externalists also assume that science has a content to be explained. Thus, . . . there is something misguided about the principles of "causality" and "symmetry" proposed by the Strong Programme in the Sociology of Knowledge. . . . These principles

24. For more detailed discussions of these difficulties with realists' uses of a causal theory of reference, see Rouse 1987b: chap. 5, and Rouse 1991a.

25. Latour 1990: 167.

> direct the student of science to explain the content of science as she would any other cultural artifact. Unfortunately, cultural artifacts are more clearly bounded in space and time than content seems to be.[26]

Undoubtedly, many of the constructivists could protest (and some already do) that the philosophical notion of content does no real work in their accounts. Their claim that the content of knowledge should be explained socially might well be equivalent to the claim that after making sense of the occasioned expressions and practices of the sciences, nothing further about scientific knowledge remains to be explained. But as I argue in Part II, this substitution is not innocuous for sociologists, because their characteristic appeal to a social explanans is conceptually linked to the idea of a content of scientific knowledge which needs explaining. The critique of 'content' thus also undermines the idea of a social constructivism.[27]

The second issue that has preoccupied philosophical and sociological interpretations of scientific knowledge has been the "ontological status" of the objects that supposedly objective knowledge represents. The notion of ontological status is closely entwined with that of the content of knowledge, but it is less readily accessible to the philosophically uninitiated. A commonsense explication of 'content' can appeal to the idea of the "thing" which is the same between two expressions that "say the same thing." But it is harder at first to make out what is philosophically at issue in the relation between representational content and what it purports to represent. The issue first arises in empiricist or instrumentalist (antirealist) criticisms of theoretical representations. Such critics, throughout the history of science, have been impressed with the utility of some theoretical representations (for example, of Copernican planetary orbits, atoms, genes, or sub-

26. Fuller 1989: 15.

27. This line of criticism has been taken up by a number of prominent contributors to sociology of science, perhaps most notably by Latour (1987) and by Michael Lynch (1985, 1992). Ironically, Woolgar (1988) has joined in criticizing the primacy of the "social," while he preserves the underlying notion of knowledge as representation. With the work of Lynch or Latour or the more recent work of Andrew Pickering in mind, one might well rewrite my historical sketch of science studies to situate this book squarely within the constructivist tradition. As noted earlier, I have chosen not to do so, to emphasize the importance of breaking with the assumptions that many social constructivists share with the philosophical tradition, which are, to repeat, indispensable to their identity as social constructivists, producing distinctively sociological interpretations. For examples of work by unrepentant social constructivists lamenting the apostasy of Lynch, Latour, Pickering, and others, see Bloor 1992 and Collins and Yearley 1992a,b.

atomic particles) while remaining skeptical of their literal accuracy. They have therefore suggested that the objects purportedly represented within such theories are human artifacts, useful as tools to predict sense experience or manipulate and control nature, but not real objects already present prior to the purposes for which they were invented. Thus, antirealists originally contrasted such pseudorepresentational constructions with genuinely representational expressions that aimed to depict real objects as they are apart from human purposes. Realists denied the contrast, insisting that all (scientific) language was representational (although not always successfully so, that is, true). For realists, scientific statements have a specific representational content, and they are true or false according to whether the world really is as they say.

Social constructivists also belong in the antirealist camp, although they draw the contrast lines between artifactual and genuinely representational discourse in a different place. All representations of nature, whether theoretical, observational, or technical/practical, are taken to be social constructions: they do not (that is, cannot) depict a world independent of human categories, practices, and social interactions but instead must be understood to be the locally determined product of those human activities. But social constructivists must thereby take the (sociological) representation of those human categories, practices, and social interactions to be genuinely representational: they describe the social structures or interactions that explain or otherwise make sense of the content of scientific representations of nature.[28] Social constructivist accounts thereby display especially clearly the close ties between the concepts of representational content and ontological status. They attribute to the objects of (natural) scientific knowledge the ontological status of a *product* of the social determination of the content of that knowledge. Latour and Woolgar forward this point with exceptional clarity: "Once the [scientific] statement

28. Here is another, related point at which the sociological tradition fractures along the lines indicated in notes 13 and 27 above. The Edinburgh Strong Programme, the Bath Empirical Relativist program, discourse analysis, and some of the early ethnographic approaches to scientific knowledge are indeed committed to taking their own (social) categories as unanalyzed within their constructivist accounts (for an especially clear and explicit defense of this commitment, see Collins and Yearley 1992a). The interesting partial exception is Latour and Woolgar (1986), who actively problematize the construction of their own account, even while (in the first edition) still using 'social' and especially 'account' as unanalyzed representations. More recently, some researchers have explicitly rejected any antirealist contrast between socially constructed "objects" and those social interests or practices represented in their explication of the processes of social construction. For critical discussion of one prominent case, Steve Woolgar's "reflexive turn," see Chaps. 8–9 below.

begins to stabilise, however, an important change takes place. *The statement becomes a split entity.* On the one hand, it is a set of words which represents a statement about an object. On the other hand, it corresponds to an object in itself which takes on a life of its own. It is as if the original statement had projected a virtual image of itself which exists outside the statement."[29]

But social constructivists are not alone in finding such a remarkably close parallel between the content of scientific statements and the character of the objects whose ontological status they assess. Scientific realists reproduce the representational content of theories when describing the objects represented because they have no other language to describe those objects than that offered by scientific theory. Realists insist, however, on the contingency of the match between theory content and object; there was no necessity that scientists would acquire theories that successfully matched the actual structure of the world. Social constructivists also see a contingency involved: there is no necessity that the social practices that constitute knowledge would stabilize sufficiently in any given case to constitute persisting (socially constructed) objects. The difference is simply in where the relevant contingencies are anchored: in the real, practice-independent character of natural objects and kinds, or in the social practices that really account for the contingency of scientific accounts of nature.

Empiricists and instrumentalists posit a more complicated symbiotic relationship between the content of scientific knowledge and the ontological status of its objects, because they typically do not take the apparent content of scientific statements at face value. An adequate theory of knowledge or theory of meaning is required to produce the requisite match between representation and object. Either the semantic content of scientific statements must be reconstructed according to a theory that accounts for what they *could* possibly represent, or their epistemic content is restricted to what knowers like us could possibly have knowledge *of*.[30] The logical positivists' verification theory of meaning illustrates the former alternative, whereas van Fraassen's constructive empiricism is perhaps the clearest example of the latter. But these more complicated cases are variations on a theme. In all three cases (realist, social constructivist, empiricist/in-

29. Latour and Woolgar 1986: 176.

30. For a useful critical discussion of epistemic and semantic alternatives to realism, see Fine 1986d. Fuller (1989: 38–40) has argued that realist and antirealist accounts of scientific knowledge can be shown to be mutually dependent, each presupposing the truth of the other in the course of explicating its own project.

strumentalist), the content of scientific knowledge and the ontological status of its objects are tightly connected notions that jointly make up a theory of representation, used to make sense of the relation between what represents and what is represented.

The role of the knowing subject within the triadic relation of representation has not been completely neglected within these traditions, even though the primary questions for their interpretations of scientific knowledge have concerned the content of knowledge and the ontological status of its objects. Postpositivist historicism, social constructivism, and their philosophical roots in Quine and Wittgenstein have directed some attention to whether scientific knowers can be treated as individual subjects or are irreducibly social as subjects of knowledge. Of concern too is whether philosophical interpretations of scientific knowledge can be restricted to the rational elements of knowers' mental activity (that is, beliefs and their inferential relations) or whether account must also be taken of their desires. And perhaps the greatest philosophical attention to the knowing subject has been directed to the relations among knowers' sense experience, their perceptual grasp of that experience, and their concepts and beliefs.

But these topics have frequently been reinterpreted to focus on representational content. Questions about how knowers perceive and observe have been translated into questions about the relation between sense data or observation sentences and other representational contents. Questions about the belief-desire nexus have similarly been reconstrued as questions about the role of "values" (the representational content of desire) in determining the content of scientific knowledge. Questions about the individual or social nature of the knowing subject have most often been worked out as whether linguistic representations acquire their content socially or individually. The result has been a rather abstract and bloodless conception of the knowing subject. Even social constructivists, enamored of the Duhem-Quine thesis and Peter Winch's Wittgenstein, have often treated knowing subjects as just networks of socially shaped beliefs and desires or as communal participants in self-enclosed forms of life.

So far, I have connected the conception of knowledge as representation that frames interpretive disputes among positivists, rationalists, realists, and social constructivists, and the conception of the legitimation project as defining what is at stake among these competing interpretations. Most participants in any one of these traditions have tried to settle questions about the legitimation of scientific knowledge by interpreting the structure or historical development of the content of scientific knowledge and the ontological status of its objects. Either the cultural authority and political autonomy of scientific inquiry have been justified by showing how the content of knowledge is determined and related to the world or challenged by showing how representations are accepted in ways that afford no global legitimation.

CRITICIZING THE LEGITIMATION PROJECT

At least four significant criticisms emerge from my discussion of the legitimation project within science studies. These criticisms address the disengagement of the legitimation project from the particular settings in which the assessment of scientific practices and claims actually matters concretely; the normative anemia of much post-1960s philosophy and sociology of science which renders it ill-suited to contribute to critical engagement with the sciences and their contributions to politics and culture; the reliance of the legitimation project on problematic notions of epistemic content, communicative transparency, and scientific community; and inadequate reflexive attention to the purposes, significance, cultural location, and epistemic and political effects of philosophical and sociological interpretations of science. Each of these points needs further elaboration.

First, the legitimation project in all its versions strains credibility by assessing the sciences globally. Accounts of the rationality, instrumental reliability, approximate truth, or social construction of scientific knowledge too often seem beside the point in the concrete situations and concerns within which the relevance, importance, or trustworthiness of scientific practices, institutions, and beliefs is most typically at issue. Consider first the selectivity of scientific investigation and the consequent question of whether the sciences actually address what is most worth knowing. Legitimating accounts of the growth of scientific knowledge, of the rationality of the methods by which that knowledge is established, or even of the approximate truth of the theories thereby arrived at usually fall silent before the question of how *significant* these achievements are.[31] Social constructivist contributions to

31. Philosophers have typically addressed this issue in two contexts. In emphasizing the virtue of explanatory power, philosophers have usually suggested that theories of greater explanatory scope were more significant and desirable than theories framed more narrowly. But apart from the difficulty of comparing the explanatory scope of different theories, the equation of significance with explanatory power raises serious problems. For example, Cartwright (1983) has argued that increased explanatory scope often requires trade-offs

the legitimation debates are little or no better off here. No one will be surprised to discover that determining which objects are most important to know about and what kinds of knowledge about them are most significant to have engages socially contested interests and interactions. Yet this recognition dissipates the rhetorical force of the counterphilosophical irony that is perhaps social constructivism's fundamental resource in the legitimation project. The significance of social contingency per se depends on contrasting accounts of scientific results as the outcome of contingent social interactions and interests rather than nature or reason.[32] Where nature and reason have no presumptive role to challenge, this general constructivist move is barren.

Similarly, the abstract and immaterial content of knowledge which philosophers put forward as rationally arrived at or approximately true is removed from any assessment that considers scientific achievements as materially and socially situated. Scientific inquiry requires material, financial, and human resources whose allocation to the sciences can be disputed. Knowledge is shaped by specific networks of socialization, communication and appraisal whose boundaries, constraints, exclusions, and deformations may be challenged on a variety of grounds. The recognition of some persons as knowledgeable differentially distributes specific forms of authority and power. Furthermore, as I argued in *Knowledge and Power,* making some things scientifically knowable requires extensive transformation of the settings in which they are to be known, accomplished by extending the materials, equipment, and practices that make it possible to reveal and monitor them.[33] Any of these practices and effects might be open to criticism even if they were arrived at by rational methods of inquiry or by developing and applying approximately true theories.

---

with phenomenological accuracy. Cartwright's arguments suggest that a more contextual assessment of the relative worth of accuracy and generality is called for. The second context in which philosophers consider significance is through the suggestion that the shared values of a scientific community determine what would be significant to investigate and which results are important achievements; presumably the rationality of the methods they employ or the approximate truth of the theories at which they arrive offers some vindication of their collective choices. Yet such *presuppositionalist* accounts of significance are of little use when it is precisely those constitutive values that are in dispute either within or outside a scientific community.

32. Woolgar (1983) has offered a lucid account of the importance of what he calls "instrumental irony" for the sociological contribution to the legitimation debates. Although this paper is widely known among sociologists of science, its challenge to sociological participation in the legitimation project has not been sufficiently assimilated within the field.

33. For a more detailed discussion of the political significance of transforming the world to make it knowable, see Rouse 1987b, esp. chaps. 4 and 7.

Another aspect of the actual contexts in which scientific institutions and practices are open to critical assessment which the legitimation debates treat inadequately is the development of knowing subjects. Philosophers and scientists often present scientific training and scientific methods not only as epistemically valuable but also as morally and politically beneficial. The cultivation of rationality, open-minded skepticism, tolerance, and cosmopolitanism have been among the elements of character that scientific work has been thought to promote. Yet other possible aspects of scientific subject formation may be at issue whose significance is not readily determinable within the terms of the legitimation debates. Questions about how scientific knowers are gendered, about the disciplines that must be internalized to become a knowing subject, or about the emotional dynamics of scientific knowers may be quite important to assessing the cultural significance of scientific practices.[34] More generally, the assessment of the exclusions and marginalizations through which some subjects are certified as knowers and accorded access to epistemically significant resources is left out of philosophical interpretations of science. For example, is there anything epistemically or politically problematic about the concentration of scientific knowers (and the resources necessary to be a successful or influential knower) within a small number of wealthy, highly industrialized nations and among particular social groups within those nations?

This disconnection of the legitimation project from concrete situations where the assessment of scientific beliefs and practices is at issue may well explain the "normative anemia" that characterizes much philosophy and sociology of science since 1970, and that is the second of my four critical themes. Philosophers have mostly lost, and sociologists perhaps never acquired, the disciplinary self-confidence needed to criticize specific scientific practices on the grounds of their own arguments and concerns. I noted above how philosophers of science have increasingly moved away from the critical stance toward scientific practice that was still a feature of positivism or Popperian falsificationism. The question underlying most historical rationalist, scientific realist, or empiricist interpretations of scientific knowledge

34. Much of the literature on gender in science (e.g., Keller 1983, Harding 1986, 1991, Haraway 1989, 1990, Traweek 1988, Code 1991) raises issues about the construction of knowing subjects, and its moral or political significance, in ways that are not effectively addressed within the terms of the legitimation debates. Michel Foucault's discussions of the power effects of disciplines and incitements to discourse are also relevant here as considerations that cannot readily be accommodated in the terms of the standard philosophical interpretations of science.

is only how best to explicate and justify the already-presumed legitimacy of scientific institutions and practices.[35] Yet, ironically, most social constructivists also avoid any significant critical engagement with scientific practices and beliefs. In part, this reflects their dissatisfaction with the philosophical interpretations of science whose normative stance is to endorse scientific practice as it is; they self-consciously try to substitute a descriptive approach to scientific practice for the philosophers' normative endorsement. Thus, for example, Collins notes in a methodological paper setting out his "empirical relativist program" that "the new sociology of science leaves pure science very nearly as it is; in the main it simply redescribes it."[36] Such an anormative descriptivism is disingenuous, of course. For one thing, redescriptions of what people do are often themselves quite consequential. For another, the intended *ironic* force of the sociological descriptions would not simply leave science as it is if we include its social status and access to resources as an important part of "pure science as it is."[37] Sociologists' commitment to an anormative descriptivism thus surprisingly reinforces their tacit allegiance to taking as their explanandum the philosophical abstraction of the representational content of knowledge. Finally, as Fuller has argued, despite the social constructivists' explicit opposition to philosophical attributions of rationality to scientists, they themselves do not hesitate to attribute to scientists a far-reaching social competence and the opportunistic rationality necessary to recognize and pursue their own interests effectively.[38]

My third criticism goes to the heart of the actual interpretations of scientific knowledge which have been proposed in fulfilling the legitimation project. I find increasingly untenable some of the fundamental concepts essential to these interpretations: the representational content of scientific knowledge, the transparent communication of that content, and the scientific communities whose identity is formed in part by their allegiance to common presuppositions.

35. Scientific realists offer a strong basis for criticism of *marginal* scientific practices and beliefs on the grounds that these rely on nonreferring concepts and false claims about the world. We may thus find realists arguing against the epistemic credentials of folk psychology, human sociobiology, or "creation science." But their arguments depend on the prior unquestioned legitimation of the referential success and approximate truth of theories in some core group of "mature" sciences.

36. Collins 1983: 98.

37. Woolgar (1983) has emphasized the importance of the difference between sociological descriptions and their intended ironic force.

38. Fuller 1992c.

Reasons for rejecting these notions will become clearer in the chapters to follow, but I will note here that my reasons focus on both the supposed immateriality and ahistoricity of epistemic content and the presumed unity and coherence of beliefs and desires within and among individual scientific knowers. Indeterminacy, instability, opacity, and difference must play a more prominent role in an adequate understanding of scientific knowledge, or so I argue.

The final criticism emerging from my reflections on the legitimation project is the reflexive inadequacy of the standard approaches to understanding scientific knowledge. In his work on the Natural Ontological Attitude, Arthur Fine argued for a reflexive incoherence in many of these interpretive approaches, one that renders their central arguments arbitrary, inexplicable, or question begging.[39] Woolgar and others have argued that reflexive applications of many sociological studies render them *rhetorically* incoherent.[40] I find Fine's arguments generally persuasive and Woolgar's deeply flawed, but I am even more concerned by a different reflexive question: How do interpretations in science studies address their own cultural and political location and the significance of their own achievements and omissions as speech *acts?* In the cases of variations on the legitimation project, my concern is that there is no way of making good sense of their cultural and political significance which would also sustain their appeal. In the case of successor projects in science studies (including my own), the question of political reflexivity is perhaps the most important and demanding challenge to be met.

SCIENTIFIC PRACTICES AND CULTURAL STUDIES OF SCIENCE

What emerges most prominently in response to these four concerns is a further articulation of the notion of scientific *practices* and a conception of interdisciplinary science studies as *cultural studies.* If the legitimation of scientific knowledge has put questions of truth and justification in the forefront, the study of scientific practices foregrounds the question of significance. The significance of scientific practices is addressed at every level of scientific research. It encom-

39. Fine 1986a,b,d, 1991, forthcoming; I discuss Fine's arguments and their implications in Chapters 2–3.

40. Extensive discussions of rhetorical reflexivity are available in Woolgar 1988a and Ashmore 1989.

passes broad, programmatic issues (which projects are worth engaging, what equipment and skills are important to acquire, what prior results must be taken into account), as well as the most mundane aspects of laboratory work (which procedural variations make a difference, which experimental runs are good ones). Questions of significance govern the codification of achievements (which results are worth publishing, how these results should be framed in an article) and the subsequent redirection of research (which recent developments in this field are important and how they shape the prospects for further work). Foregrounding significance reminds us that most truths about the world are scientifically irrelevant or uninteresting; recognizing the difference between important and insignificant claims is indispensable for understanding scientific practice. Discussing significance also shows how the critical assessment of scientific work is multidimensional, extending well beyond the justification of claims to truth or empirical adequacy.

Questions of significance can be addressed, I argue, only in the context of ongoing scientific practices. Practices are patterns of activity in response to a situation. Practices are dynamic, because these patterns exist only through being continually reproduced. Their coherence and continuity thus depend both on coordination among multiple participants and things and on the maintenance of that coordination over time. Because scientific practitioners are geographically dispersed, responsive to local opportunities and constraints, imperfectly communicative, discontinuously policed, and differentially located within fields of overlapping scientific practices, there is room for considerable slippage in the ongoing reproduction of the "same" practices. Furthermore, practices are intrinsically open to multiple interpretations (where the most basic way to "interpret" a practice is to continue it or extend it to new situations,[41] because the reproduction of the practice cannot be determined by rules. In any case, scientific practices typically encompass considerable interpretive differences (both explicit and unarticulated), even synchronically. As a result, scientific practices have a complex temporality. As Hans-Jörg Rheinberger tellingly summarized the temporal slippage in the ongoing reproduction of scientific practices, "The recent is the result of something which did not happen, [whereas] the past is the trace of something which will not have occurred."[42]

41. This conception of interpretation as practical is developed more extensively in Rouse 1987b: chap. 3.
42. Rheinberger 1994: 67.

Scientific practices thus embody a continual tension between intelligibility and incoherence. Dispersion and openness to multiple interpretations continually threaten their coherence, but because any specific research only becomes intelligible and significant as a contribution to a project shared with others, a relentless pressure also exists to adjust one's work to fit in with what others are doing. This ongoing tension is manifest in what I call the narrative reconstruction of science. By 'narrative' I do not mean a literary form but a way of comprehending the temporality of one's own actions in their very enactment. Scientists make sense of what they are doing by understanding it as a response to the situation presented by past research and an anticipation of future developments. That is, they continually enact a narrative in the midst of which their present activities (and those of others) would be intelligibly situated. Their narrative comprehension must be continually reconstructed to accommodate the recalcitrance of other people and things and to appropriate the new possibilities thus made available. As a result, the sense and significance of what scientists do is continually reconfigured by the subsequent course of research. Understanding scientific practice as narrative reconstruction shows how scientific work becomes coherent and significant without having to rely on problematic notions of scientific communities, consensus, or background content. What scientists share is not a background of beliefs but a situation that they may belong to and understand in partially divergent ways.

This account of the narrative reconstruction of scientific practices points toward the dynamic character of scientific understanding and communication. My discussion of the dynamics of scientific knowing is modeled on Thomas Wartenburg's situated, dynamic account of power.[43] Wartenburg argued that power is mediated by "social alignments": one agent's actions effectively exercise power over another only to the extent that other agents' actions are appropriately aligned with the actions of the dominant agent. Power relations are dynamic, for they depend on sustaining these alignments over time and in response to subordinate agents' efforts to resist or bypass them. Knowing is similarly mediated by "epistemic alignments"; skills, models, concepts, and statements become informative about their objects only when other people and things interact in constructive alignment with them. The analogy requires some modification of Wartenburg's discussion of power, because he mistakenly restricts the mediation of

43. Wartenburg 1990.

power to social alignments of human agents. As I argued in *Knowledge and Power*,[44] reconfiguring the material and practical setting of action also constitutes or transforms power relations. But once the heterogeneity of power alignments is acknowledged, we can recognize analogous alignments that situate and mediate the practices of assessing, attributing, relying on, or contesting knowledge.

Taking the dynamics of these epistemic practices seriously encourages a deflationary or nonreifying account of scientific knowledge. A deflationary account of knowledge is directly opposed to those epistemological views that take knowledge to constitute a theoretically coherent kind. A conception of *scientific* knowledge as a coherent kind has been implicitly presumed by most participants in the legitimation project; it only makes sense to claim that scientific knowledge as a whole is approximately true, rationally arrived at, or socially constructed and interest-relative if there is such a (kind of) thing. But deflationary accounts also have no truck with skepticism. Skepticism depends on the identification of knowledge as a theoretically coherent kind; only then can the skeptic claim that that kind is empty.[45] The deflationist, by contrast, recognizes a wide range of examples of knowledge but denies that they collectively constitute a coherent kind. Perhaps surprisingly, deflationary accounts should also be distinguished from epistemological eliminativism. Eliminativists argue that our ordinary talk about knowledge should be replaced by some more informative vocabulary with which epistemological categories are incommensurable (in Chapter 7, I take as salient examples Paul Churchland's neurophysiological eliminativism and Fuller's sociological eliminativism).[46] Deflationists take the ordinary uses of the term 'know', and the associated practices of justification, reliance, and criticism, to be valuable and generally unproblematic, while denying that they can be explicated as instances of a well-formed kind. The contrast to eliminativism is especially interesting, because deflationary accounts permit more heterogeneity in what is potentially relevant to ordinary ascriptions of knowledge in particular contexts than is permissible in eliminativists' proposed replacement theories.

44. Rouse 1987b: chaps. 1, 7.

45. Williams (1991) offers a clear and cogent account of the difference between a deflationary account and either skepticism or epistemological foundationalism. In developing this account, he argues for a sense of 'foundationalism' that includes coherence theories of knowledge. On Williams's account, foundationalists and skeptics agree in being "epistemological realists," that is, in taking knowledge and its various subdivisions (knowledge of the external world, of other minds, etc.) to be coherent and unified kinds.

46. Churchland 1989; Fuller 1989.

My treatment of the dynamics of "epistemic alignments" is not itself another general account of knowledge which would conflict with a deflationary understanding because it does not characterize knowledge as a whole or as a theoretically coherent kind. The concept of an epistemic alignment is too open-ended to serve in this way as an epistemological theory. There is no reason to believe that present practices have exhausted (or present theories have exhaustively described) the sorts of elements that might contribute to epistemic alignments or the ways in which they might effectively align. Nor do they constrain what someday might count as informative alignments. One of the historically important features of the sciences has been their role in changing what can count as knowledge. What justified such changes has not been an analysis of how they more adequately exemplified what genuine knowledge really is (such analyses, when offered, are better understood as codifications of the conceptual changes such reconfigurations accomplished). Their "justification" has instead been their realization of more informative ways of interconnecting what people do with their environment; but the purposes for which they count as "informative" are intelligible in turn only through the maintenance of appropriate configurations of practices.

One of the striking consequences of this deflationary account of knowledge is its undoing of any *general* category of representation. The seemingly natural idea that knowledge is representational accounts for much of the intuitive appeal of the legitimation project. This idea motivates the question of how representations could ever be appropriately connected to objects represented. Whether representations are located in people's heads, sentences, skills and practices, or forms of life, the question takes the same general form. Similarly, objects may be construed as things in themselves or as things perceptually, practically, or conceptually manifest without escaping the representationalist problematic.

I try to undermine the seeming obviousness of representationalism in several ways. First, a deflationary account of knowledge makes the epistemological side of representationalism otiose. It allows us to account for the significance and intelligibility of scientific practices and other inquiries and for the defeasible ascription of 'knowledge' without postulating any representations in which knowledge is located. Second, the notion of a practice is articulated more carefully to avoid reinscribing representationalism covertly. Many of the increasingly common philosophical and sociological appeals to "social practices" still reproduce representationalism by equating practices with rules, norms, or skills that knowers or agents embody or enact.[47] Practices thereby become analogues to the knowing subject and are consequently distinct from the objects practiced with or on. As an alternative, I suggest conceiving of practices as meaningful situations or configurations of the world. Agents or knowers do not bestow meaning on practices but instead they count as agents and knowers only through their place in ongoing patterns of practice. This conception of practices eventually needs to be developed as a full-scale account of intentionality; my more limited aim is to indicate the advantages of such a conception for understanding and engaging with the sciences.

Perhaps the most telling challenge to representationalist accounts of science arises from criticism of their underlying semantic intuitions. Everyone who studies the sciences recognizes the centrality of language for scientific understanding. Philosophers and sociologists of science usually then presume without much argument that language works representationally, and that presumption makes representationalist conceptions of knowledge seem both natural and unavoidable. But alternatives exist. Donald Davidson, for example, has worked out an attractive and plausible account of meaning and communication which does not reify reference relations, conventional meanings, or even natural languages.[48] Davidson's discussion of communicative practice meshes well with a deflationary account of scientific knowledge and with my proposal to model science studies on cultural studies. It would require a very different book to defend seriously the correctness of Davidson's approach to semantics. I do claim, however, that the availability of such an account undercuts an important motivation for representationalism and the legitimation project and encourages attention to alternative approaches.

What implications does my account have for the relation of scientific practices to philosophical, historical, sociological, or ethnographic studies of science? The various participants in the legitimation project attempt to make sense of the sciences from a standpoint that I call "epistemic sovereignty." Epistemic sovereignty would permit a de-

47. Turner (1994) provides telling criticisms of the most familiar versions of such appeals to "social practices" within social theory and philosophy. Although he does not talk specifically about "representationalist" conceptions of practice, as I would, it is clear that his principal targets are those views for which the term 'practice' is interchangeable with such terms as "tradition, tacit knowledge, *Weltanschauung*, paradigm, ideology, framework, and presupposition" (2).

48. Davidson 1984, 1986.

tached interpretation and assessment of conflicting claims and methods. Its standpoint is that of the impartial referee or judge, and it postulates a sharp distinction between science and the metascience practiced in science studies.[49] Ideally, scientific practice would aim to settle truth claims, whereas science studies would address the metalevel methodological or sociological issues that would explain and perhaps justify those scientific determinations. My approach, by contrast, argues that natural science, like natural language, includes its own metapractice. The point is not to reduce science studies to science narrowly construed but to expand the conception of "science" to include critical reflection on its own practices and goals. I mean both that scientific practice inevitably includes such reflection (often unthematically) and that science studies should be understood as engaging with scientific practice rather than just interpreting it. In Fine's elegant summary, "There is no legitimate hermeneutical *account* of science, but only a hermeneutical activity that is a lively part of science itself."[50]

A long and prominent tradition in epistemology and political philosophy insists that epistemic and political criticism can be justified only from a standpoint of sovereignty. Those who give up that standpoint supposedly have no basis for holding anyone's epistemic or political practices accountable to standards not of their own devising. Within science studies, however, it is the defenders of epistemic sovereignty who now find it difficult to hold the sciences critically accountable. There are several aspects to my attempt to reinvigorate the critical dimension of science studies without assuming epistemic sovereignty. I have argued elsewhere that neither epistemic nor political criticism is disabled by rejecting a standpoint of sovereignty.[51] My notion of 'epistemic sovereignty' is explicitly formulated by analogy to Michel Foucault's insistence on a political criticism that would "break free . . . of the theoretical privilege of law and sovereignty."[52] My more general arguments for the possibility of a nonsovereign critical standpoint were formulated explicitly as a reflection on Foucault's discussions of power/knowledge and on critics such as Jürgen

49. For a discussion of the notion of "epistemic sovereignty," see Rouse 1993, forthcoming. As I argue in the latter paper, the relativist stance frequently assumed by social constructivists is a clear case of epistemic assessment from a sovereign standpoint: it differs from most philosophical legitimations of science only in what counts as an appropriately detached standpoint and in the alleged consequences of adopting such a standpoint.

50. Fine 1986c: 148 n. 9.

51. Rouse 1993, forthcoming a.

52. Foucault 1978: 90.

Habermas, Nancy Fraser, Charles Taylor, and Richard Rorty, who have argued that his work is either deeply conservative or normatively confused.[53]

These general arguments have been exemplified by many feminist studies of science, which develop a critical practice that does not assume a standpoint of sovereignty.[54] In this respect, feminist studies of science stand in sharp contrast to the social constructivist tradition.[55] Both traditions insist on the ineliminably social and political dimensions of the formulation, assessment, and certification of scientific knowledge. Social constructivists, as we have seen, infer from the social contingencies of knowledge production a skeptical or relativist commitment to the legitimation project. Knowledge can be justified only within particular social contexts, and there are no transcendent standards for assessing its rationality or verisimilitude. Relativism is unacceptable to feminists, however, for multiple reasons. Feminists cannot accept that sexist beliefs and practices are justifiable relative to sexist forms of life. Feminists also frequently argue that feminism as a social movement and a political practice can *improve* scientific practice and understanding. Finally, feminists often recognize the pretension to epistemic sovereignty as itself a disablingly gendered position whose masculine bias needs to be exposed and transcended.

Feminist science studies can be readily understood in terms of the narrative reconstruction of science and the formation of epistemic alignments and counteralignments. Feminist criticism is not an attempt to impose a politics of gender on the sciences from the "outside." Feminists have instead shown how gendered assumptions and practices have been integral to the shaping of what counts as significant and reliable scientific work. Much of the initial work in feminist science studies originated within the sciences, both in the sense that the work was accomplished by feminist scientists and in the sense that it criticized specific constructions of gender within biology and psychology.[56] Subsequent feminist work situated within history, philosophy, and sociology or anthropology of science may have tried to

53. Habermas 1986, Fraser 1989, Taylor 1986, Rorty 1982, 1989.

54. Prominent examples of feminist science criticism which explicitly address the kinds of epistemological issues I consider here include Haraway 1989, 1990; Longino 1989; Harding 1991; Keller 1985; Wylie 1991, 1992, forthcoming; Nelson 1990; and Code 1991.

55. Rouse (forthcoming b) provides an explicit articulation of the contrast between social constructivist and feminist conceptions of the social and political dimensions of knowledge.

56. Prominent examples include Bleier 1984; Fausto-Sterling 1985, Birke 1986, and Keller 1985. Haraway (1989) also discusses explicitly feminist themes that are developed in many behavioral studies of primates in the 1970s and 1980s.

formulate more general theoretical criticisms of gendered science, but these projects have not lost touch with more specific engagements with scientific practice. Feminist epistemic practice has attempted to realign scientific practices both by contesting scientific work directly and by constructing feminist counteralignments that can resist attempts to suppress or ignore feminist criticism within the sciences and science studies. Unlike social constructivists, who have challenged the cultural authority of the sciences without contesting the specific practices or claims recognized as authoritative, feminist critics have tried to *participate* in the practices through which knowledge of the world is constructed as authoritative.

The contrast between feminist and social constructivist studies of science can be generalized to suggest a better model for the interpretive and critical practices of science studies. I argue that the philosophy of science, and science studies more generally, should be reconceived as interdisciplinary cultural studies of science. Cultural studies focus on the emergence of meaning within human practices. The focus on practice and meaning do not, however, oppose culture to nature or the physical world. Material culture also belongs within "cultural" studies, including the ways in which the world resists or reinforces people's efforts to make sense of it. The distinctions between culture and nature or between meaningful action and its meaningless physical environment thus belong to the field of meanings which cultural studies address instead of constituting the field's boundaries.

Cultural studies also reject any standpoint of epistemic or political sovereignty. Culture and cultural meaning are not objects of study from which the inquirer can be detached but refer to the field of meaning within which cultural studies themselves are situated along with their proximate objects of study. The practice of cultural studies must consequently be reflexive and politically engaged. The interdisciplinary orientation of cultural studies appropriately reflects the heterogeneous contributions to the formation of culture. Admittedly, many of the most familiar examples of cultural studies have focused on literature and popular culture.[57] Their principal theoretical tools have been adapted from political economy, social theory, interpretive anthropology, history, and post-Hegelian philosophy. But cultural studies of science must be committed to a critical understanding of

57. Grossberg, Nelson, and Treichler (1992) and During (1993) provide wide-ranging surveys of current approaches and achievements within cultural studies.

scientific practice and culture and not to a previously agreed-on constellation of theories and disciplines. My aim in proposing cultural studies as a model for science studies is therefore not an ironic disciplinary imperialism.[58] I do hope to encourage new ways to think about scientific practices and achievements and to promote a much more expansive conception of philosophy, sociology, or history of science. But it cuts both ways to recognize that the culture of science and technology is virtually coextensive with culture at large. Cultural studies more generally should not and will not remain unaffected by more sustained attention to scientific and technological practices and to the interpretation and reconstruction of nature as integral to other cultural and political projects.

THE STRUCTURE OF THE BOOK

In Part I of this book I develop the interpretation and criticism of the legitimation project and its underlying assumptions, thereby pointing toward some of the more constructive themes of Part II. Chapter 1, "Philosophy of Science and the Persistent Narratives of Modernity," contains my most extensive discussion of the continuity between philosophy or sociology of science and the debates over the legitimation of "modernity." Although I discuss positivism, realism, and social constructivism, my central focus is the "piecemeal" tradition that emphasizes the historically unfolding rationality or progress of science as the ground for its epistemic and cultural legitimation. The reason to discuss Laudan, Shapere, Miller, and Galison so extensively is that they claim to respond to many of the philosophical problems that I attribute to the legitimation project. By emphasizing the continuity between their approach and the positivist, realist, and constructivist traditions, I indicate that the shared and problematic assumptions of the legitimation project require rather more far reaching philosophical reconception than their more familiar orientations would suggest.

I also introduce the question of the relation between contemporary philosophy of science and Vienna Circle positivism, one to which I return in Chapter 4, "Should Philosophy of Science Be Postpositivist?" Realists, rationalists, and constructivists frequently tout their own

58. "Ironic" because cultural studies have frequently challenged the constraints imposed on inquiry and criticism by established disciplinary models and standards.

respective approaches as finally revealing and avoiding the debilitating flaws of positivism, seeing the competing traditions as failing to break with the underlying positivist errors. In Chapter 1 I suggest that each tradition in turn fails to make such a radical break with positivism. Chapter 4 offers a different twist to the question of vestigial positivism. First, I show how their preoccupation with finally becoming postpositivist has made it difficult for philosophers of science to recognize and respond constructively to some of the most promising work in science studies, which draws on the resources of feminist theory and interdisciplinary cultural studies. Second, I show how some of the best work within philosophy and sociology of science displays important constructive affinities with positivism. This discussion of positivist themes in the work of Nancy Cartwright, Arthur Fine, Ian Hacking, and Bruno Latour will also foreshadow some important themes of Part II: the importance of scientific practice, the intelligibility of practices, the openness and heterogeneity of the sciences, and the engagement of philosophy of science with scientific practice and the culture of science.

The two intervening chapters of Part I are critical reflections on a series of important articles that Fine published in the 1980s.[59] Fine's work has important continuities with my own critique of the legitimation project within philosophy and sociology of science. As an alternative to the problematically shared stance that realists, empiricists, constructivists, and piecemeal rationalists take toward science, Fine proposes the "Natural Ontological Attitude" (NOA). In "The Politics of Postmodern Philosophy of Science," I explore what NOA amounts to, with special attention to its stance toward cultural and political criticism of scientific practices. Fine's proposed attitude combines a thoroughgoing antiessentialism about the sciences ("NOA thinks of science as an historical entity, growing and changing under various internal and external pressures")[60] with an insistence that the sciences do not need an overall interpretation or legitimation. These points might suggest that NOA forecloses cultural and political criticism of the sciences along with realist, instrumentalist, or constructivist interpretations. Taking several feminist studies as test cases, I argue that the best reading of Fine's proposals leaves ample room for historically situated critical reflection on the sciences. These arguments also point toward an important theme in Part II: serious political engage-

59. Fine 1986a,b,d, 1991, forthcoming.
60. Fine 1986d: 172.

ment with scientific practice requires not a social constructivist version of the legitimation project, which might undercut wholesale any scientific pretensions to epistemic and cultural authority, but rather a historically and culturally situated political criticism.

Chapter 3, "Arguing for the Natural Ontological Attitude," considers what sort of arguments are needed to make Fine's proposals plausible. The minimalist justification that Fine favors simply holds NOA up for consideration, appealing to interlocutors' judgment and good sense. This approach is Fine's bow to positivism; he admires the Vienna Circle's hostility to metaphysics and extends it to encompass epistemology and semantics, but he hopes to substitute his natural ontological attitude for their self-defeatingly circular theories of meaning. Fine's approach nevertheless fails, both philosophically and rhetorically. Philosophically, this approach generates yet another self-defeating circularity, one comparable with those Fine ascribes to the various participants in the legitimation project. Rhetorically, it underestimates the cultural entrenchment of the legitimation project, the widely felt *need* for authoritative underwriting of our epistemic and political commitments. The chapter strongly suggests the need for my more robust defense in Part II of a deflationary philosophical understanding of 'truth', 'meaning', 'justification', and 'knowledge'.

In the first part of this book I do not pretend to offer a direct or comprehensive criticism of the various philosophical and sociological contributions to the legitimation project. Such has already been accomplished for the most part within the available literature in philosophy and social studies of science. The proponents of the dominant traditions have often insightfully grasped the shortcomings of competitors' approaches and of their own as well. The most fundamental difficulties confronting scientific realism, for example, have been effectively articulated in well-known work by Laudan, Fine, and van Fraassen.[61] The subsequent proliferation of modest, moderate, internal, entity, and other high qualified invocations of the word 'realism' strongly suggests the intractability of the difficulties confronting a robust realist contribution to the legitimation project.[62] Van Fraassen's resurrection of empiricism has also not fared well in the hands of critics, despite widespread admiration for the philosophical virtues of his exposition and defense: constructive empiricism today is more

61. Laudan 1981; Fine 1986a, 1986d; van Fraassen 1980.

62. Fine (1991) effectively extends the familiar criticisms of a robust scientific realism to some of its most prominent attenuated versions.

a challenging foil for realists than a widely accepted alternative interpretation of scientific inquiry and achievements.[63] The social constructivist tradition has often been seriously misconstrued or oversimplified by would-be philosophical critics, but now some quite widely recognized unresolved difficulties also confront the standard approaches to the "sociology of scientific knowledge," many of which have arisen within the sociological tradition itself.[64]

Rather than rehearse once more these well-known interpretations and arguments, Part I aims to redirect the scholarly response to them. I hope to convince readers of the need for different answers to different questions about the sciences rather than a better version (or better defense) of realism, empiricism, social constructivism, or the historically unfolding rationality of science. The chapters here thus offer diagnoses of shared assumptions underlying the familiar difficulties; they show why some supposed alternatives to the standard approaches remain committed to the same assumptions or encounter parallel difficulties. Above all, they highlight themes and problems within the legitimation project which point toward my own concerns in the second half of the book.

In Part II I work out my alternative to the legitimation project within science studies, beginning in Chapter 5 by discussing "the significance of scientific practices." In *Knowledge and Power* I already proposed that the sciences should be understood primarily as practices. After briefly recapitulating some of the important themes from those arguments, I more carefully differentiate that claim from other, more recent philosophical and sociological discussions of practice. Some philosophers appeal to practices in their quest for a non-Cartesian conception of human subjectivity or intentionality, emphasizing publicly accountable agency instead of mental representation. They understand intentionality in terms of practices so as to distinguish more adequately the realms of freedom and nature or the domains of the human and the natural sciences. Social constructivists

63. Many of the classic challenges are collected in Churchland and Hooker 1985; Fine 1986b, 1986d also deserve mention as especially telling critiques. Even those inclined to defend a thoroughgoing empiricism (e.g., Mitchell 1988) recognize that the position needs to be substantially reformulated to meet familiar objections.

64. Barrett and Roth 1991 and many essays in Brown 1984 exemplify the extensive, mostly philosophical, literature that tendentiously criticizes the sociology of scientific knowledge on the basis of highly unsympathetic or misleading accounts of the sociological project and its claims. Among the more telling assessments of classical sociology of scientific knowledge (SSK) by judicious critics are Latour 1992; Woolgar 1983, 1992; Lynch 1992; Proctor 1991; Fine forthcoming; Roth 1987; and Hesse 1986.

similarly accentuate human agency when they take *social* practices to constitute what can count as nature or as scientific knowledge. Neither approach is consonant with my interpretation of science as practice. I understand practices not as the doings of human agents but as the meaningful situations within which those doings can be significant. Meaningful patterns are not bestowed on the world by agents or their shared forms of life but emerge from patterns of interaction within the world. My understanding of scientific practices is therefore part of a conception of meaning that eschews both naturalistic reductions of intentionality to causal processes and humanistic exaltations of subjectivity and freedom.

In the following chapter, "Narrative Reconstruction, Epistemic Significance, and the Temporality of Scientific Practices," I further explore the conception of practices and agency developed in Chapter 5 by describing scientific work as narrative reconstruction. This chapter indicates why scientific practices should be understood as situated within contested narrative fields and introduces the idea that the sense of scientific practices and claims is thereby temporally deferred. In developing these themes, I emphasize the importance of the question of the significance and intelligibility of scientific practices, argue against postpositivist conceptions of scientific communities and their shared presuppositions, and suggest an alternative to familiar conceptions of the unity of science. The chapter concludes with my initial discussion of cultural studies as a model for interdisciplinary studies of scientific knowledge.

Chapter 7, "The Dynamics of Scientific Knowing: Understanding Science without Reifying Knowledge," is in many ways the pivotal chapter in Part II. In this chapter, I develop the conception of scientific knowledge which results from my understanding of scientific practices. My account of scientific practices cannot just be assimilated to a more familiar epistemological holism that would incorporate practices within "background knowledge." I argue for this claim first by modeling the dynamic and disseminated character of scientific knowledge on Wartenburg's account of the dynamic and situated alignments that mediate social power. This analogy shows how 'knowledge' is not some*thing* that knowers possess or exchange but instead characterizes how they are situated in the world. I then show that the epistemological consequences of this way of understanding knowledge are best captured by a deflationary conception of scientific knowledge. A deflationary move shifts attention from knowledge as a coherent kind of thing to the practices of assessing authority, justification, reliability, and informativeness. It thereby shows the inappropriateness of the legitimation project and the attempts to explain or account for scientific knowledge in general.

I take up the significance for science studies of a Davidsonian approach to language and communication in Chapter 8, "Against Representation: Davidsonian Semantics and Cultural Studies of Science." Davidson offers a dynamic understanding of language which parallels the dynamic conceptions of knowledge and power in the preceding chapter, utilizing a deflationary, semantic conception of truth which is the model for my deflationary approach to scientific knowledge. I make three principal uses of the Davidsonian approach: it shows the dispensability of the conceptions of linguistic representation which both motivate and support the legitimation project; it enriches and extends my conception of scientific practices as dynamically engaged with the world; and it further develops the notion of science studies as cultural studies by offering a framework for understanding theoretical modeling and experimental manipulations as discursive practices. Davidsonian accounts of interpretation might initially seem to conflict with the critical dimension of cultural studies. Cultural studies often attend to cultural difference and the power effects of distortion, silencing, and incommensurability through which difference diffracts, whereas Davidsonians deny even the possibility of radical difference in conceptual schemes. The chapter concludes by showing this conflict to be merely apparent. Davidsonian semantics can actually be an instructive resource for understanding and criticizing the effects of power in linguistic understanding and communication. The final chapter, "What Are Cultural Studies of Science?" offers a more extensive account of my conception of cultural studies as a critical alternative to the social constructivist tradition and in so doing draws together many of the earlier themes of the book. Social constructivism is the foil for this programmatic statement, because for all its difficulties, it is still the richest and most informative of the traditions that take up the legitimation project, and many of its historical and ethnographic studies must be taken into account by any adequate approach to interpreting the sciences. My argument self-consciously draws on others' work that I take to exemplify cultural studies of science. My conception of cultural studies is not put forward as a strikingly new way to think about the sciences but as an attempt to formulate a more philosophically and politically satisfying interpretation of the more recent achievements of constructivist sociology, contextualist history, and feminist theory of science.

Part I

RETHINKING THE TRADITIONS: POSITIVISM, SCIENTIFIC REALISM, HISTORICAL RATIONALISM, AND SOCIAL CONSTRUCTIVISM

[1]

Philosophy of Science and the Persistent Narratives of Modernity

Philosophers of science in recent years have become increasingly dissatisfied with the legacy bequeathed by the successively central traditions of logical empiricism, postempiricist historicism, and convergent realism. Despite the fundamental conflicts among participants in these philosophical traditions, many critics have been struck by what they have in common.[1] All three traditions provide a unified narrative structure within which to write the internal or philosophical history of science. All three assume that fundamental methodological concepts for the sciences (for example, explanation, confirmation, and reduction/translation) are in important respects invariant across scientific disciplines and their historical development. All insist on a unitary and hierarchical relation (in one direction or the other) between theory and observation or experiment. Finally, and perhaps most important, all three traditions find that a general theory of meaning and/or reference proves indispensable to understanding science philosophically.

Many of those who identify and reject these characteristic features of the tradition from positivism to realism also propose alternative approaches to the philosophy of science which share many important features. The philosophical programs offered by Miller, Laudan,

1. For example, see Miller 1987; Laudan 1984; Shapere 1984; Galison 1988; and Fine 1986c,d.

Shapere, and Galison all forego appeal to general accounts of meaning or reference in favor of domain- or topic-specific principles of methodology. Within those various scientific disciplines or domains, they insist that methods are not a given and fixed context for scientific work but are themselves developed and improved in the course of research. The result is some form of methodological bootstrapping and overlapping periodization of the history of science, as hierarchical accounts of the relation between theory and experiment or among theory, methods, and cognitive values are replaced by models of "reticulation" or "intercalation." In the end, these "piecemealist" critics (to adopt Shapere's label) find it important to preserve the idea of a distinctive internal history of science and to advocate preserving the cultural autonomy of science which that internal history supposedly ideally reflects. But typically, they salvage the internal/external distinction by preserving one or two of the traditional internalist categories (progress, rationality, reality, truth) at the cost of decoupling them from the others.

Galison provocatively formulates the common objections to the tripartite legacy of positivism, postempiricism, and realism by linking it to debates in other disciplines over modernism and postmodernism as cultural narratives. He argues that logical positivism has deep and important affinities to the flourishing of high modernism in various cultural domains between the two world wars and that the postempiricist alternatives offered by Kuhn, Feyerabend, Lakatos, or the early Laudan do not make a fundamental break with the positivists' modernism. And although Galison does not discuss convergent scientific realism, its most prominent proponents readily fit within his story of more recent philosophy of science as the unfolding of a modernist discourse. An important part of Galison's project is to situate his own approach to the history and philosophy of science with respect to the philosophical tradition and the broader cultural debates over modernity: he is not afraid to call his own view of the sciences a "critical 'postmodern' model."[2]

One might initially question the usefulness of Galison's suggestion that we situate philosophy of science within the slippery context of the debates over modernity. The vocabulary of "modernity" has become increasingly problematic within philosophy, literary theory, and social thought, as theorists have speculated about the possibility and significance of being or becoming "postmodern." For one thing, the princi-

2. Galison 1988: 207.

pal features and indicators of modernism or modernity look strikingly different from the perspectives of the arts and architecture, politics, technology and science, or social theory. Also, the discussion of the high modernism that flourished between the world wars is often conflated with reflection on the larger narrative strategies that have constructed the "modern world" or the "West" as an object of historical knowledge and cultural identification. But still deeper differences exist over what the alleged differences between modernity and postmodernity are supposed to represent. For some, postmodernity seems to represent a world historical transformation, as three or four centuries of more or less continuous "modernization" come to an end, leaving us on the threshold of new and portentous possibilities; for others, postmodernism represents a veneer that poorly disguises more of the same, either an antimodern conservatism or a celebration of capitalist consumer culture.[3] Still others take postmodernity to be less a development within the narrative periodization of culture than a skepticism toward the very project of periodizing modernity, such that "we learn with great delight that there is no such thing as a modern world."[4]

In another way, however, these controversies over the significance and extent of "modernism" can be an asset. Although only Galison self-consciously adopts the vocabulary of modernity, Miller, Laudan, and Shapere join him in taking themselves to have made a significant break with the traditional assumptions in the philosophy of science, to a great extent the assumptions that Galison identifies with modernism. But it is an interesting question to what extent these various philosophical programs actually do abandon the discourse and narrative strategies of modernity. To that extent, Galison's formulation, with all the complexity and controversy it invokes, may be a useful probe to explore just how significant a shift in direction is actually achieved by their various challenges to the tradition that extends from positivism to scientific realism. In any case, whatever it contributes to the philosophy of science, Galison's appropriation of the discourse of modernity addresses a glaring deficiency in much of the philosophical, literary, and social scientific discussion of modernity and postmodernism. For despite the obvious centrality of natural science and the interpretation of science for the narrative construction of "modernity" and despite the real affinities between much of the philosophy

3. Lyotard 1984, Habermas 1982, Jameson 1984.
4. Latour 1988: 201; for a more extensive discussion, see Latour 1993.

of science and the narrative strategies employed in that construction, surprisingly little sophisticated attention has been given to the significance of the philosophy of science for understanding what to make of modernity and postmodernism.

In this chapter I will use Galison's discussion of modernism in the philosophy of science to address the foregoing concerns. I will begin with a more careful exploration of how philosophy of science has participated in the narrative strategies and tropes that we now associate with the construction of "modernity," which will enable me to ask how we should understand the philosophical programs put forward by Miller, Shapere, Laudan, and Galison. I take up the vocabulary of modernity rather agnostically. I want to remain uncommitted for now about many of the controversies I have alluded to above. But I employ their vocabulary in part as a tool to examine whether and to what extent these critics have actually broken with the legacy of positivism, postempiricism, and scientific realism, in part to explore the continuity between philosophical interpretations of science and other projects of cultural interpretation for which talk of modernity and postmodernism is now a commonplace, and ultimately to ask what it might take to make a fundamental break with the assumptions embodied in the modernist discourse of the philosophy of science, if these critics have not yet done so successfully.

Galison unequivocally takes the high modernism between the two world wars as the context within which to interpret contemporary philosophy of science. He argues that logical positivism "had close affinities with other brands of modernism; it paralleled the manifestoes of art, architecture, and politics of the 1920's."[5] On his account, despite their obvious points of conflict with the positivist tradition, the early postpositivists (Kuhn, Feyerabend, N. R. Hanson, Lakatos) shared fundamental modernist assumptions with logical positivism. What the positivists and their critics supposedly had in common was the project of constructing a "master narrative" of grand scope, which projected on the history of science a unity "enforced by the provision of a privileged vantage point."[6] The accounts of science they offer invoke a hierarchical and reductionist relation between theory and experiment which was based on a general account of the structure of linguistic meaning and reference. In the positivists' case, Galison explicitly analogizes their account of the role of protocol sentences to

5. Galison 1988: 201.
6. Galison 1988: 201.

the Bauhaus's reduction of architecture to combinations of elementary geometrical forms. The postempiricists replaced this account with a theory-dominant holism, which was likewise grounded in a general theory of meaning. In place of these grand unifying, hierarchical, linguistically grounded narrative strategies, Galison offers as the beginnings of a "postmodern" central metaphor the picture of relatively autonomous and intercalated levels of development of theory, experiment, and the material culture of instrumentation.

Even if one follows Galison in making high modernism central to understanding the tradition in the philosophy of science, one finds difficulties with his interpretation. For his parallelism between the positivists and their early critics to work smoothly, it is important to place at the center of the positivist program their empiricist account of the reduction of theoretical sentences to protocol sentences. Otherwise, it would seem implausible to insist (as Galison does) that a hierarchical relation between theory and experiment is the key modernist feature of both programs. Yet some historical work suggests that one misunderstands logical positivism if one places a Humean empiricist reductionism at its forefront. Michael Friedman has convincingly argued that we must take seriously the neo-Kantian roots of the Vienna Circle and recognize the centrality of their affinities to Hilbert's metamathematical formalism.[7] On Friedman's interpretation, the crucial problem in the positivist tradition was how to account for mathematical truth once the development of polyadic logic and relativistic physics made Kantian appeals to intuition both unnecessary and untenable. Ironically, both Moritz Schlick and Rudolf Carnap began with holistic, theory-dominant accounts of perceptual judgment and were only gradually driven toward empiricism by the felt need to defend the cultural primacy of mathematics and mathematical physics in the face of the severe problems of theoretical underdetermination which their formalism engendered.

If one accepts the broad outlines of Friedman's interpretation of the positivist tradition, its continuity with modernism is actually enhanced, although the shape of that continuity differs from Galison's account. Galison's analogy between protocol sentences and elementary geometrical forms was somewhat strained: the analogy loses the latter's mathematical formalism and its austere restraint to a *limited* repertoire of elementary shapes. By contrast, Friedman's interpretation of positivism offers multidimensional affinities with modernism.

7. Friedman 1984, 1987.

Modernist movements in the arts were characteristically formalist and stressed the separation of formal structure from referentiality. In this respect, it is crucial to recognize with Friedman that, despite its reputation, Carnap's *Aufbau* was not a phenomenalist tract but a general theory of the formal construction of concepts. As Carnap himself stressed: "*Each scientific statement can in principle be so transformed that it is nothing but a structure statement.* But this transformation is not only possible, it is imperative. For science wants to speak about what is objective, and whatever does not belong to structure but to the material (i.e., anything that can be pointed out in a concrete ostensive definition) is, in the final analysis, subjective."[8] Like modernist movements in painting, architecture, music, and literature, the positivists' views of mathematics and science emphasized disciplinary purity, with mathematics severed from Kantian intuition and unified science conceived as the application of formal scientific methods to the domain of what is objective (that is, to the domain of formally structured concepts). And of course, as Galison rightly emphasized, the positivist manifestoes were militant tracts for a rationalist, progressive internationalism which "resemble far more the daring pronouncements of the Italian Futurists, or the Russian Constructivists, than they do the more usual philosophical fare."[9]

Despite these useful parallels, however, I believe one must situate more recent philosophy of science not just within the context of twentieth-century modernist movements in the arts but also within the larger discourse of modernity. Important aspects of the positivists' project, carried on by their successors, make sense within larger modernist narrative strategies but do not show up within the narrow frame of reference selected by Galison. Among those features of positivism which escape the analogy with artistic modernism are the importance of rationality and method, of unified science as objective knowledge, and of the primacy of mathematics and mathematical physics within the whole of culture. Without taking into account these more broadly modernist features of positivist philosophy of science, one might well find the inclusion of the sciences within Ernst Cassirer's account of symbolic forms a better fit with modernism than Carnap's or Hans Reichenbach's positivism.

To see this argument, we must say more about what is meant by the discourse of "modernity," of which the high modernism cited by Galison is supposed to be a significant component. Recall that I want

8. Carnap 1928: sec. 16.
9. Galison 1988: 201.

to give "modernity" the broadest plausible interpretation in order to see which putative features of modernism are significant for the more recent tradition in philosophy of science and whether these are continued in the work of Miller, Shapere, Laudan, and Galison, despite their aim to break with the most basic assumptions of that tradition. Recall also that I want to remain noncommittal about the truth of the narratives of modernity; after all, even if the modernist self-understanding is in important respects false, it has nevertheless been influential. If nothing else, "modernity" represents a widespread cultural self-understanding.

I thus take "modernity" to be characteristically the name for a narrative strategy. It describes the historical construction of the modern world, a world that differs from all predecessor cultures in certain characteristic and systematically linked ways. Among the aspects that are frequently cited as distinctively modern are the following:

(1) Secularization—not simply the decline of religious practice and belief but also the separation of various realms of human life from theology and the privatization of religious and other fundamental matters of concern into individual beliefs and commitments

(2) The constitution of "humanity" (the unity of abstractly differentiated *individuals*) as the *subject* of representation and knowledge, source of all values, and possessor of rights and moral dignity

(3) The development of distinct domains of knowledge and practice (for example, law, the economy, science) whose autonomy is institutionally recognized and protected

(4) Rationalization—formal procedures of calculation developed more intensively, extended to more and more domains of human practice, and unified across domains

(5) The rapid growth of science and technology as the quintessentially modern human practices and the correlated understanding of nature as inert object of knowledge (science is perhaps the only human activity for which the appellation *modern* has almost no controversial force)[10]

(6) The expansion and concentration of productive resources (usually identified with the beginnings of capitalism as a mode of production)

(7) The extension of European ("modern") culture across the globe (colonialism and postcolonial modernization/Westernization)

(8) The self-referential narrative legitimation of modernity itself as the progressive realization of freedom and truth

10. I am grateful to Thomas Ryckman, who brought this point to my attention in responding to an early version of this chapter.

It is not difficult to recognize how tightly linked these various features of modernist narratives are, such that their separation into distinct aspects of modernity is somewhat artificial. The constitution of humanity as knowing subject and source of value replaces theology in a secularized culture, transforming theological and political traditions into matters of individual belief and desire. What escapes such privatization are purely formal (rationalized), procedural rights. Such rights are attributed to the rational, distanced subject free to choose among explicit possibilities. Secularized domains of knowledge and practice are also constituted through rationalization, which is taken to be the characteristic form of scientific method and economic behavior. Science and technology provide the paradigm cases of secularized domain formation and are increasingly the basis of economic expansion and political or cultural Westernization. The expansion of Westernized culture both appeals to and is taken to justify the modernist account of universal human values and forms of knowledge. More generally, all modernist narratives of legitimation appeal to those values to justify further secularization, rationalization, economic expansion, and the allocation of resources and authority to scientific research and technological development.

When one thinks of the narrative legitimation of modernity, one typically thinks of stories of progress which portray modernization as the advance of knowledge, the establishment of human rights and humane practice, political democratization, the creation of wealth, and the control of nature, all to the benefit of humanity. Such narratives have always been accompanied by antimodern counternarratives, however. Antimodernist narratives typically have the same global sweep and a structure similar to that of stories of modern progress. Indeed, they tend to be the same story, with the valorization reversed. The same practices are now described as the destruction of rootedness and community, the reduction of anything of worth to merely instrumental calculation or arbitrary preference, the displacement of wisdom and judgment by bureaucratic rules, disastrous inattention to or suppression of holistic interconnection across domains, the domination and despoilation of nature, and the shallowness of consumer culture. We shall take *modernity* to encompass the tropes and strategies of both modernist and antimodern narratives and the felt need by the proponents of each to forestall the appeal of the other.

We can now ask to what extent positivist philosophy of science and its successors belong to this broader tradition of modernist narrative.

It should not be difficult to see how the logical positivist program, for all its technical sophistication, is very much at home within this narrative field. The positivists' hostility to metaphysics was a continuation of the Enlightenment diatribes against theological superstition: they took metaphysics to be the successor subject to theology. Central to their views was the cultural primacy of mathematics and science, exemplified by mathematical physics. Science was identified with the employment of formal method, for the sake of instrumental success. Derivatively, *science* connoted an attitude of tolerant skepticism. The practice of scientific method and the adoption of its progressive attitude promised the accumulation of instrumental capability at the service of humanity, and the ruthless detection and elimination of error. Their program of Unified Science was both the technical consolidation and purification of the domain of objective knowledge constituted by formal rationality and the assertion of universal, internationalist, and progressive values. Their version of empiricism emphasized the universal aspects of individual experience. As has been widely noted, the political spectrum of the Vienna Circle from liberal social democracy to scientific Marxism was culturally of a piece with their formalist philosophical project. Perhaps less widely noted, their advocacy of a skeptical and tolerant scientific attitude admirably began at home: most of the telling philosophical criticisms that eventually ruined the positivist program were generated internally.

The gradual collapse of the positivist philosophical program in the 1950s and 1960s brought to prominence a new set of themes and problems in the philosophy of science. The reception of postpositivist philosophy of science in the larger culture, commonly associated with the names of Kuhn and Feyerabend, often emphasized antimodernist themes: dethroning the sciences from their cultural preeminence, associating scientific practice with dogmatic or irrational belief and particularistic communities, and replacing formal methods and rational self-criticism with the language of religious conversion and unanalyzable holistic gestalt switches. This tendency itself suggests a move within the dialectic of modern and antimodern narratives. But it also misses the much greater prominence of positively modernist themes within the work of Kuhn and Feyerabend themselves and also of such well-known postempiricists as Imre Lakatos, Stephen Toulmin, Mary Hesse, and Larry Laudan.

The widespread abandonment of formalism within postempiricist philosophy of science reflected a tension immanent to modernist narrative. Modernism celebrates both the epistemic autonomy of increasingly specialized domains of knowledge and social practice and their unification by formal procedures of reasoning and justification. The positivists subordinated the claims of disciplinary autonomy within the sciences to formal method and the unity of science; their postpositivist critics largely abandoned formalism for the sake of the autonomy of disciplines and research programs within disciplines. Despite their turn against formalism, the postpositivists mostly retained an intense concern for an account of metascientific rationality (best exemplified by Lakatos's all but vacuous account of degenerate research programs, Laudan's desperate hope to measure problem-solving effectiveness, and everyone else's obsession with the supposed irrationalism disclosed by Kuhn and Feyerabend).[11] Such concern for rationality was closely connected to rationalization in the sense prominent in the narratives of modernity: this concern was part of the project of distinguishing the "internal" history of science, constituted by factors distinctively relevant to science as a domain of inquiry, from those "external" influences that must be banished in the name of Galileo against the Roman Church.

Despite its widely reputed demise, the modernist association of science with progress flourished in early postpositivist philosophy of science. Lakatos, Laudan, and the others strove valiantly to preserve a more or less traditional conception of progressively cumulative empirical knowledge. Although Kuhn and Feyerabend are notorious for having rejected this enterprise, ironically they did so by promoting different conceptions of progress on quintessentially modernist grounds. Feyerabend's insistence that we should prefer progress to method appealed explicitly to Mill's arguments for the epistemic utility of tolerance. Kuhn offered a conception of scientific progress which tried to eliminate its teleological components, while preserving the idea that scientific work was distinctively progressive.

Galison rightly pointed to the focus on high theory in postpositivist philosophy as an important vestige of modernism, but he subtly misunderstood its significance because he overemphasized the empiricist element in logical positivism. Galison took the turn toward theory as simply an inversion of the positivists' foundationalist hierarchy. This interpretation is mistaken on two counts: as I already noted, following Friedman, the positivists were much more sympathetic to holism than Galison acknowledges; more important, Galison's reading ignores the extent to which the postpositivists' appeal to a highly abstract and

11. Lakatos 1978; Laudan 1977. A representative account of Kuhn and Feyerabend as irrationalists is Scheffler 1982.

general notion of theory supplanted some of the positivists' accounts of formal methods. The substantive commitments about the natural world embodied in a theoretical worldview or research program were supposed to determine which evidence was relevant to the epistemic assessment of particular claims and how much weight that evidence carried; they offered exemplary patterns of legitimate explanation, and they provided the context within which both theoretical and observational terms acquired meaning. In short, a network of theoretical beliefs was supposed to do the work of qualitative and quantitative formal analyses of confirmation, deductive-nomological accounts of explanation, and reduction sentences.[12] This approach involved the replacement of purely formal relations between sentences by substantive theoretical commitments but was accomplished at a high level of abstraction. As Ian Hacking somewhat uncharitably put it, "The word 'theory', a word best reserved for some fairly specific body of speculation or propositions with a fairly definite subject matter [has been used] to denote all sorts of inchoate, implicit, or imputed beliefs."[13] It is no accident that Joseph Sneed and his followers were tempted to formalize Kuhn's concept of a paradigm or that others have described this as a "network" model of theory and evidence. As Galison rightly suggested, despite their historical studies, early postpositivist philosophers relied more on a general theory of meaning allied (via Frege) to that of the positivists than they did on close and detailed readings of the interdependence of substantive theoretical claims in particular cases. In the end, it seems reasonable to claim that the modernist penchant for abstraction and formalism was not abandoned with the demise of positivism; only its focus was shifted, from first-order deductive logic and the quest for a formal account of induction to a fairly undifferentiated treatment of theoretical networks.

The ascendancy of postpositivist accounts of research programs was short-lived. What happened to modernist narrative then? If one respects disciplinary boundaries, early postpositivism was supplanted by scientific realism; if one looks more broadly at global interpretations of science responding to the postpositivist tradition, sociological constructivism offered an alternative track. Both approaches have nevertheless worked very much within the narrative frame of modernity.

At first glance, scientific realism seems to break with fundamental

12. Insufficient attention has been paid to the fact that Kuhn's famous concept of a "paradigm" was explicitly proposed as an alternative to correspondence rules as an interpretation of the connection between theory and world; see Kuhn 1970: chap. 5.

13. Hacking 1983: 175.

aspects of the narratives of modernity. By restoring to our image of science the quest for metaphysical truth, realism seems to turn away from the modernist emphasis on rationalization and instrumental success. Realists also restore the connection between representation and reference and thereby look to the world rather than to the knowing subject as the key to understanding scientific knowledge. Van Fraassen effectively satirized this apparent turn away from modernity by likening the realist's principal arguments to Aquinas's Five Ways for proving the existence of God.[14] A closer look at the main arguments for realism suggests that this appearance of a return to a premodernist understanding of science is misleading. Contemporary arguments for realism depend on taking science to aim, successfully, at increased instrumental capability.[15] The approximate truth of scientific theories is justified as the best explanation of this presumed instrumental success. Furthermore, the most sophisticated realist arguments presuppose the earlier postpositivist account of the determination of observation and methods by global theories; the inference from instrumental success to realism is mediated by this overarching role of theory.[16] Finally, the turn away from the subject is belied by the fact that contemporary realism depends on a general verificationist move within the treatment of reference, a move that allows one to write a realist narrative of the history of science: realists apply a principle of charity to interpret what previous scientists' terms should *count* as referring to, based on the interpreter's present understanding of what kinds of objects the world contains. Realism thereby offers a robust modernist narrative of the history of science as a story of progress. The antiscientific superstition that succeeds theology as the realists' modernist bête noire is not metaphysics, which they take to be respectably scientific, but a priori epistemology, which they insist should be naturalized in turn.

Constructivist sociology in its various guises may also seem initially promising as an interpretation of science which avoids the modernist tropes, but once again appearances mislead. Constructivist rhetoric often oscillates uneasily between a hypermodernist critique of rationalist superstition and an antimodernism that too readily recapitulates the modernist narratives in opposition. In the first guise, sociologists have chided philosophers for an apriorism that avoids scientific (sociological) research, uncritically taking scientists' self-

14. Van Fraassen 1980: chap. 7.

15. Fine 1986d makes this feature of realist arguments central to his challenges to realism, for example, in his "Metatheorem 1."

16. For a good example, see Boyd 1984.

justifications at face value and appealing to mysterious influences by "ideas" or "reasons" instead of a hardheaded, austere restriction to local, material causes. Their history of science is also a history of Enlightenment, in which sociologically naïve scientists and philosophers of science replace tyrants and priests as the purveyors of superstition to be overcome.

In their antimodernist guise, the sociologists reject the narratives of scientific progress, only to retell the story as one of irrational, ideological domination, as the sciences gradually acquire an unjustified cultural hegemony. The causal powers shaping the narrative of the history of science are shifted to what most philosophers supposed to be "external" to science, but the split between internal and external factors is basically preserved: what is "rational," "epistemic," or "cognitive" is still opposed to what is "social," even if the former categories are now taken to be empty. Scientists work to serve their interests (whether as scientists or as members of a particular social class) or to negotiate resource relationships; however conceptualized, the picture typically presented is a characteristically antimodernist one of rationalization without rationality. In the end, most social constructivists agree with the philosophical tradition that the autonomy and cultural authority of the sciences "in the modern world" call for the global legitimation offered by modernist narratives of progress; they deny only that the need can ever be legitimately satisfied.

This uneasy oscillation within the sociological challenge to philosophical hegemony in interpreting science—between modernist legitimation of the sociological standpoint and antimodern debunking of progress and the cultural primacy of science—is evident in the cultural relativism that is the most widely discussed feature of the various constructivist positions. Relativism is a paradoxically modern affliction. It seems to reject the primacy of science, the universalism of "modern" Western practices and beliefs, and the concomitant legitimation of progress. Yet it does so by taking up a distanced, representational stance toward the totality of human beliefs and practices, affirming the closure of the "worldviews" of other human subjects (whether identified with individuals or with particular cultures), and insisting on an attitude of respect or tolerance for such views, one comparable to Kant's or Mill's outlook toward rational subjects. And inevitably, the metaworldview of the relativist, if not taken to be true, is at least put forward as morally or methodologically progressive. My point is not the usual one that relativism is self-contradictory and hence false but that it is an incoherent cultural stance that leaves its proponents clearly within a narrative tradition, though unstably so (which does not by itself show that such a position cannot or should not be occupied).

This all too hasty consideration of how more recent philosophical reflection on science has moved within the protean narrative field of "modernity" should by now raise an insistent question: Why not take up such modernist narrative within philosophy of science? Is it even possible to avoid doing so (in which case the objection would at best be pointless)? It is with these questions in the background that I want to examine the place of Miller, Shapere, Laudan, and Galison with respect to these narratives that have shaped the tradition from which they try to break.

Galison only indirectly indicates the argumentative rationale for his identification of fundamental assumptions shared by logical positivism and its postpositivist critics as modernist, but the message he intends is clear: modernist narratives of legitimation have lost credibility. Their repeated manifestation in many guises in multiple contexts, and the evident failures that they have often seemed thereby to produce, make them look increasingly like artifacts of the way disciplines and practices are constructed or written about. The narratives of modernity appear to impose a spurious unity on irreducibly diverse practices and beliefs. Furthermore, the specific versions of "modernity" have been repeatedly criticized as epistemically, politically, and aesthetically inadequate or oppressive. For all the analytical deficiencies of Lyotard's essay on the postmodern condition, his diagnosis of the present intellectual condition as one of growing "incredulity toward metanarratives" seems quite accurate.

This chapter is not the place to try to summarize the various complaints put forward in many contexts about the ways the more general constructions of modernity have come to seem inadequate. Rather, to connect it to a critical assessment of the work of Miller, Shapere, Laudan, and Galison, I would like to show how the motivation for this incredulity has arisen within philosophy of science. I want to begin with a question: Why does it seem important to so many philosophers to disclose common assumptions between logical positivism and many of the postpositivist schools in philosophy of science? To be sure, positivism has been widely discredited, but that was accomplished by direct and detailed arguments against specific theses advanced by positivists, not by analogy to other discredited views.[17] Why

17. The exception is the extent to which this description characterizes the objection that the positivists' critiques of their opponents' forays into "metaphysics" were self-referentially inconsistent, because their own view was unverifiable, hence metaphysical.

not proceed simply by comparably direct argument against postpositivist, realist, and constructivist approaches?

To answer this question, we must consider the significance of the widespread rejection of logical positivism within Anglo-American philosophy of science. The positivist tradition provided the context for most philosophical work (in English) in the philosophy of science for some three or four decades. It engaged philosophers in technical issues of considerable subtlety and sophistication and shaped the discussion of science in other fields. To a considerable extent, the positivist tradition did not seem to be just an influential position within philosophy of science: it was the philosophy of science. Against this background, the extent of its collapse was shocking. Within a very short time, it passed from this preeminent status to that of a view whose deficiencies were so glaring that it was difficult to see how it could ever have been taken so seriously. The commitment to formalism in methodology appeared in retrospect to have turned philosophical attention away from science itself and toward logical puzzles disconnected from any insight into the practice or success of science. Indeed, the widespread judgment of hindsight was that positivism offered a model of science which no actual science had ever even approximated and which would be disastrous if prescribed as an ideal. Its technical sophistication seemed to many a clear case of powerful tools grossly misdirected, while its foundationalist account of scientific observation seemed naïve and uncritical. Its fundamental outstanding problems (confirmation and theory reduction) appeared to present obviously insoluble difficulties. This harsh rejection undoubtedly reflected unfairly a retrospective reinterpretation and judgment. To a significant extent, the image of positivism which underlies this response does not do justice to the positivists' texts (although a more sensitive reading is unlikely to make Vienna Circle positivism a live option again). But the overwhelming impression made on succeeding generations of philosophers looking at science was that the positivists were simply unable to see their supposed object of interest—scientific knowledge—because of the concepts and assumptions they brought with them and imposed on the task of a philosophy of science. The lesson to be learned from the hold that positivism had had on the philosophical imagination was that only a fundamental break with the positivists' assumptions would enable philosophers to see something other than the reflection of their own commitments.

Despite an apparent consensus among philosophers of science that the discipline must make such a break, there has been no consensus about exactly what the mistaken assumptions were. So each successive line of interpretation presents itself as the one that finally grasps the full measure of what went wrong with the positivist program, whereas its predecessors/competitors unwittingly committed the same mistakes that had led philosophers so far astray for so long. The revenge that positivism exacted on its predecessors was (and is) the felt need to break decisively with positivism, which inevitably led to a continuing preoccupation with the positivist program (or at least its reinterpretation as a foil for later views).

This rhetorical stratagem was certainly present in the early postempiricists' rejection of formal methods and foundationalism in favor of historicism and theoretical holism; it also showed up with a vengeance in scientific realists' turn from Fregean to causal theories of reference and their revival of concern for the truth of scientific theories. Not surprisingly, this stratagem remains central to the latest generation of philosophical rebels. Richard Miller begins his 1987 book with this proclamation: "In a broad sense of the term, positivism remains the dominant philosophy of science. And the most urgent task in the philosophy of science is to develop a replacement for positivism that provides the guidance, philosophical and practical, that positivism promised."[18] Laudan, whose first philosophical efforts followed closely in the footsteps of Kuhnian postpositivism, now insists that "it has been insufficiently noted just how partial Kuhn's break with positivism is, so far as cognitive goals and values are concerned."[19] Dudley Shapere repeatedly emphasizes the "complementarity" between the insights and difficulties of the positivists' views and those of Kuhn and Feyerabend.[20] Meanwhile, he insists of the philosophical underpinnings of scientific realism that "the Kripke-Putnam approach to the interpretation of science in terms of reference, and the opposing [positivist and postpositivist] approach in terms of meaning, share fundamental assumptions, and despite their deep differences, those approaches stem from the same tradition."[21] We have seen, of course, that Galison identifies the shared assumptions underlying previous work in the philosophy of science with modernism, which he hopes finally to surpass with his critical postmodern model of scientific change.

18. Miller 1987: 3.
19. Laudan 1984: 70.
20. Shapere 1984: xiii–xviii, 156–66.
21. Shapere 1984: 384–85.

How do these critics conceptualize their allegedly fundamental breaks with a unified tradition in the philosophy of science? Allowing for differences in formulation and emphasis, the piecemealists identify four ways in which their own interpretive approaches reorient the philosophy of science away from the debilitating assumptions of the tradition from positivism to realism. First, all claim to reject what Miller called the "worship of generality" on methodological issues in favor of principles which are "topic-specific" or domain-determined and which change with advances in scientific understanding. Second, they reject hierarchical determination in favor of multidirectional relations among theories, methods, values or goals, instrumentation, and experiment or observation. Third, all reject "the supposition that the nature of science can be illuminated by an examination of alleged necessities of language which are independent of the results and methods of scientific inquiry."[22] Finally, all aspire to underwrite the story of the history of science as a narrative of progress, of rational inquiry, and/or of the gradual approach toward truth, a task that they argue was rendered untenable when set in the context of positivism, Kuhnian postpositivism, or convergent realism.

There is much to be said about the detailed variations on these themes, the sundry arguments and examples provided in support of them, and other distinctive features of these four philosophers' views. In this chapter, however, I am painting with a broader brush. My concern is to ask both how fundamental a break these four proposals would be with the tradition stemming from positivism and to what extent these breaks have actually been achieved. It is here, I believe, that my extension and reconstruction of Galison's appeal to the broader cultural narratives of modernism may be useful in identifying an underlying continuity between these critics and the tradition they hope to have put behind them.

The obvious place to start is with that aspect of the narrative strategies of modernity which was never part of the piecemealists' self-conceptualized breach with the positivist tradition: the narrative justification of progress, rationality, and/or truth. No longer is it assumed that the rational pursuit of inquiry readily progresses and that progress takes us toward truth; this package of categories is dismantled to preserve one or more of the pieces. Thus Shapere's philosophical project is to show how the history of science constitutes domain-specific forms of rationality, as scientists learn how to learn about

22. Shapere 1984: 383.

nature and gradually to distinguish the space of relevant (internal) reasons from considerations external to their domain of inquiry. As methods and results arrived at within a domain are gradually freed from specific, relevant doubts, no good reason exists not to call them true, but the hero of the narrative is domain-specific rationality.

Like Shapere, Miller characterizes the overcoming of specific, locally relevant doubts as the progressive element in the history of science, but he changes the emphasis. The history of science can, he argues, be characterized to a significant extent in realist terms, as the appeal to local, topic-specific truisms can show that many scientific existence claims have been established beyond any reasonable doubt and that theories about those entities are approximately true. Rationality (in the local, innocuous form of reasonable belief) follows accordingly.

Laudan vociferously rejects realism but insisted all the more strongly in 1984 that "we need a single, unified theory of rationality" which can account for the progressive growth of consensus in the natural sciences.[23] His reticulational model of justification was supposed to provide this theory. Three years later, he appeared to recant the preeminence of rationality, taking progress to be the crucial meta-methodological concept: "We can readily make claims about the progress of science, even if we know nothing whatever about the aims or rationality of earlier scientists . . . because, unlike rationality, progress need not be an agent-specific notion."[24] Meanwhile, Galison remains more cautious in evaluating the history of science on a large scale, as befits his objection to modernist narrative. His account of how experiments end quietly eschews the terminology of rationality and truth.[25] Nevertheless, in the end he proposes a more complex periodization of the history of science than are provided by the modernist models he criticizes, and he notes the effect on our assessment of that history: "In an interesting way, the brick model may suggest how the practice of physics can involve discontinuities at so many levels and yet not disintegrate. . . . [Bricks stacked] one directly on top of another . . . would leave the wall all too vulnerable to shock along the alignment. Only in an intercalated arrangement do the bricks give the edifice strength."[26]

All these narratives of modern scientific progress proceed by way

23. Laudan 1984: 3.
24. Laudan 1987: 28.
25. Galison 1987.
26. Galison 1988: 210.

of an "internal history" of science: to this extent, they remain the heirs of the Enlightenment critique of theological superstition and the positivist attack on metaphysics. Like the postpositivists they oppose, they have largely ceded the formal unity of scientific methods in order better to promote the disciplinary closure and autonomy that are the other side of modern rationalization. As Steve Fuller trenchantly pointed out, when "Shapere grounds the enhanced epistemic power of modern science in its increasingly systematic and self-contained reason-giving practices, he is apparently oblivious to the fact that they are just one instance of the rationalized accounting procedures [for example, economic and legal] that have generally emerged in response to the complexity of modern society."[27] The construction of autonomous domains of knowledge and practice which can be rationally administered in accordance with internal goals and standards is a fundamental part of familiar stories of "modernity."

In constructing their accounts of the rationalization of scientific disciplines, the piecemealist critics have not abandoned invariant hierarchical organization within the structure they attribute to science: they have only adopted oligarchy as the preferred form of hierarchy. Galison demands equal rights for theory, experiment, and the material culture of instrumentation, whereas Laudan charts the equilateral interaction among theories, methods, and cognitive values. Shapere shows a complexity to the relations between theory and observation which neither positivists nor postpositivists recognize. But all insist that these factors must be the crucial ones in understanding the epistemic or cognitive development of scientific knowledge. Such factors as communication networks and forms, institutions, resources, or material culture (the last of which Galison does acknowledge) undoubtedly influence what scientists actually say and do, but they are conceptually distinct from the methodological factors that govern (or ideally govern) scientific practice, and they allegedly make no contribution (or only inadvertent contributions) to the progress, rationality, and/or truth of scientific achievements. Shapere claims that these exclusions have been learned in the course of constructing scientific domains, whereas Laudan says only that they could be justified by a naturalistic theory of methodology.[28] Miller does not discuss them at all, perhaps presuming the appropriateness of excluding all such fac-

27. Fuller 1989: 20.
28. Specifically, in Laudan 1987.

tors by leaving them out of his accounts of how some existence claims were established beyond reasonable doubt.

Unfortunately for the piecemealists, one cannot both maintain the idea that there is a coherent internal history of science and also insist that the philosophy of science should not appeal to a general theory of language as the basis for interpreting the sciences. The connection between the idea of an internal history of science and a general theory of language shows up in several considerations raised by Fuller.[29] He noted that the very idea of an internal history of science requires a distinction between the *content* of scientific theories, methodological principles, observations, reasoning, and cognitive values and the *context* within which this content is conceived, communicated, and established. For a history of the content of science conceived in this broad way to be intelligible, it must be possible for the *same* content to be transmitted and preserved through varying formulations in different contexts. This claim suggests an underlying theory of meaning which would justify the identification of this common content across various modes of presentation to different intended (and unintended) audiences, whose goals and assumptions may differ.[30] If, like Shapere, you see the internal history of science as a convergence on relevant patterns of reasoning within domains, then you need to account not just for common content denoted by various inscriptions and utterances but also for shared (or at least convergent) *reasons* for uttering and accepting that content. This requires that reasoning be sufficiently transparent to be identified and agreed upon. To those who think that the requisite account of meaning or content is innocuous or uncontroversial, Fuller's nominalist demurral should be informative. He offers these "homely observations":

> Knowledge exists only through its embodiment in linguistic and other social practices. These practices, in turn, exist only by being reproduced from context to context, which occurs only by the con-

29. Fuller 1989, esp. chaps. 1–2.

30. This objection may be more specific to the piecemealists than Fuller's formulation suggests. Distinctions between content and context are invoked (often implicitly) on various occasions, for different purposes, with different boundaries of demarcation; such ordinary attributions of content do not, I believe, presuppose a general account of language or language use. But the piecemealists want to *fix* the boundary between the content of scientific domains, disciplines, or topics and the contexts that remain external to this gradually demarcated content, as the ground for distinguishing internal and external history of science. It is this reification of content which I believe must rely on a general theory of meaning. I am grateful to Thomas Ryckman for pointing out the need to distinguish my position from Fuller's in this way.

> tinual adaptation of knowledge to social circumstances. However, there are few systemic checks for the mutual coherence of the various local adaptations. If there are no objections, then everything is presumed to be fine. Given these basic truths about the nature of knowledge transmission, it follows that it is highly unlikely that anything as purportedly uniform as a mind-set, a worldview, or even a proposition could persist through repeated transmissions in time and space.[31]

One may well argue in defense of internal history against Fuller's linguistic nominalism (or against a more moderate pragmatism that treats all content/context distinctions as occasional), but it is then hard to claim that the resulting realism about meaning or content does not reproduce the "sense of the linguistic nature of the science problem that underlay both the positivist credo of unification and the antipositivist departures from it."[32]

The piecemealists' professed disenchantment with the positivist and postpositivist penchant for generality likewise promises more than it delivers. Shapere proposes a turn away from general theories of rationality toward accounts of reasoning within specified domains (or disciplinary subfields) of inquiry. But his account of the rationalization of domains of inquiry is itself a general narrative pattern within which to understand all the history of science, which he takes to be "a process of gradual discovery, sharpening, and organization of relevance-relations, and hence a gradual separation of the objects of [a domain's] investigations and what is directly relevant from what is irrelevant thereto: a gradual demarcation, that is, of the scientific from the unscientific."[33] Miller similarly insists that arguments for or against realism cannot be resolved at a general level but must appeal to "topic-specific truisms." But the role of such topic-specific truisms in scientific reasoning about unobservables is not itself topic-specific, for that would then undercut Miller's realism. Arthur Fine aptly summarizes the difficulty:

> The conclusion the realist is after is that it would be unreasonable (i.e., irrational) not to believe (say, in molecules), for only this will defeat the instrumentalist. . . . Thus Miller's analysis of the case invokes the following perfectly general principle of rationality: *if one*

31. Fuller 1989: 4.
32. Galison 1988: 206.
33. Shapere 1986: 6.

> *has grounds for belief, and no specific doubts on the matter, then it would be irrational not to believe.* . . . [This invocation of principle] is surprising in Miller. For Miller is a particularist, and the passion for this sort of generality in our treatment of a topic like rationality is among the things that Miller identifies as the legacy of logical positivism.[34]

Galison and Laudan are more straightforward about their own generalist stance. Galison proposes his "postmodern model" as a "central metaphor" for understanding the history of science, no less general or metaphorical than its positivist and postpositivist predecessors, but superior on its merits. Laudan adopts an explicitly normative metamethodological stance from which he happily assesses the history of science as a story of progress with respect to our present cognitive goals and interests, which need not coincide with the goals and interests of the various agents whose work contributed to that story. In the end, I believe that Shapere's and Miller's histories also amount to Whiggish stories of progress, a claim requiring argument at greater length. In any case, they all invoke, as the narrator of their philosophical histories, the rational subject who chooses among fully explicit possibilities for belief and action.

What can we conclude about the piecemealists' attempts to set themselves apart from the tradition from positivism to scientific realism? Despite many significant differences from that tradition, their interpretations of science belong firmly within it when it is understood as the project of situating the sciences within the philosophical narratives of "modernity." The history of science is interpreted as a legitimating narrative of progress, rationality, or the successful pursuit of truth, without which scientific practices and results would be arbitrary or unfounded.[35] The need to tell such a story about science (or at least most science) requires a backing off from the rejection of the "worship of generality" or Shapere's "Inviolability Thesis," for the construction and the elaboration of autonomous domains of specific reasons become the general and inviolable features of scientific development. In these latest versions of modernist legitimation, the pursuit of knowledge is rationalized within domains by the exclusion of "ex-

34. Fine 1991: 89.

35. It is not simply the project of praising science which is characteristically modern but rather the opposition between narrative legitimation and arbitrariness or groundlessness, such that science is felt to be *in need* of such legitimation. Those who take the successful critique of philosophical legitimations of science as evidence for its arbitrariness are no less modernist than the philosophical tradition.

ternal" concerns, whether their externality is established by the imposition of present goals on the narrative of progress (Laudan) or is taken to be the outcome of the narrative construction internal to that domain (Shapere).[36] Within domains, "science" is still identified with the construction and transformation of disembodied, decontextualized representations, explicitly assessed and chosen by a rational subject. The positivists' rational reconstruction of formal relations between sentences and the postpositivists' abstract and sweeping theories, research programs, and worldviews have been abandoned. But their replacements (background information, chains of reasoning, topic-specific truisms, and the like) are no less abstracted from the variety and complexity of interrelated utterances and inscriptions that occur within scientific work. As a result, the philosophy of language, broadly construed, remains crucial to the piecemealists' philosophy of science, despite their refusal to rely on standard theories of meaning or reference. They postulate a history of the *content* of scientific reasoning and belief which is abstractable from its multifarious local instantiations over time.

Suppose I am right that we should firmly identify piecemealist philosophy of science with the tradition that runs from positivism to scientific realism, which the piecemealists take themselves to have repudiated in important ways. This does not by itself show their views to be untenable, although it does undercut their rhetorical strategy of making their case as the first or only real alternative to positivism. In the end, I believe that their narratives of legitimation are no more credible than those they aim to replace, although much more argument would be needed to establish that claim. But my interpretation does raise the question of what it would take to break out of the modernist tradition in the philosophy of science, if such is even possible. For if this is not possible, it is certainly no fair criticism of the piecemealists that they, too, have failed to put modernity behind them.

I offer the following suggestions in a skeptical spirit, for any such proposal is an invitation to treat it the way I have treated the piecemealists, as a target for assimilation to the tradition. These suggestions reflect not a principled program but an amalgamation of observations on the difficulties encountered by the piecemealists and

36. The piecemealists thus follow the early postpositivists against the positivists in taking the rationalization of inquiry within autonomous domains to be more fundamental to science than its formalization.

on various approaches to thinking about the sciences which strike me as promising alternatives. Think of it as an interesting but untried recipe that would still take much skillful adaptation to come out well, if it can do so at all.[37]

The grand narrative legitimation of the history of science as a history of rationality, progress, or the search for truth must go by the wayside, but it must be accompanied by the debunking of science which too often motivates the repudiation of such modernist narratives. Take legitimations of scientific practices and beliefs always to be partial and to take place in specific contexts, for particular purposes, to which large-scale (de-) legitimation has little relevance. Partly this would involve a more skeptical attitude toward the concept of science itself. High-energy physics, meteorology, molecular genetics, natural history, petrochemistry, and cosmology are interestingly different from and complexly related to one another (nor do difference and complexity vanish *within* such domain boundaries). In any case, the boundaries of "science" cannot be readily located by identifying such domains: technology and material culture,[38] metrology and the creation and extension of phenomena,[39] the recruitment of allies and resources,[40] science classes, museums, advertisements, and popular culture[41] may be as intimately a part of "science" as are theories, observations, and the reasoning that connects them. The idea that there is a "natural world" for natural science to be about, entirely distinct from the ways human beings as knowers and agents interact with it, must similarly be abandoned.[42]

Do not think of knowers as featureless and abstract reasoners but as situated agents with an inescapably partial position.[43] Instead of thinking of the sciences in terms of "representations" of the world, look at the actions they involve and the ways they transform the situation for further action.[44] Insofar as reports of scientific results represent something, these representations are embodied,[45] and they represent only by being *used* for some purposes. Where they appear,

37. For the image of the recipe as a model of knowledge, I am indebted to Fraser 1988 and Heldke 1988.

38. Ackermann 1985, Latour 1987, Galison 1987.

39. Hacking 1983, Rouse 1987b, Latour 1987.

40. Latour 1987.

41. Haraway 1989, 1990: chap. 10.

42. See Rouse 1987b: chaps. 6–7.

43. Haraway 1990: chap. 9.

44. Rouse 1987b.

45. Fuller 1989.

how they are constructed, who is to read them, and how they circulate are inextricable from what they say.[46] Even the repetitions of sentences and the reproduction of phenomena and machines are translations that affect what they say and do.[47] Accounts of the content of scientific work, the stuff of internal history, are themselves further translations, ones that must be treated like their scientific counterparts by asking what they do. There is no occupation of a metastandpoint from which to interpret science which is not a move within the contested terrain where scientists themselves operate, a terrain in which "science" names both the outcome and the shape of the contested terrain.[48]

Would such a program for thinking about science help take us beyond modernity, or even "modernity"? That remains to be seen, as does whether that aspiration is worthwhile. If it does take us beyond modernist narratives, it does not get beyond telling stories, in which "science" names a character(s). The hope is that, in contrast to the *Bildungsromane* of "modernity," the stories we may thereby learn to tell still hold out some possibility of being believed and perhaps even of being true.

46. Rouse 1987b: chap. 4.
47. Latour 1987.
48. Rouse 1987b: chap. 6; Fine 1986b; Latour 1987.

[2]

The Politics of Postmodern Philosophy of Science

Since the heyday of Vienna Circle positivism in the 1920s and 1930s, the philosophy of science has been thoroughly intertwined with what it is now fashionable to call the politics of modernity. For the Vienna Circle itself, the parallels to modernist movements in other domains of culture are very strong. Peter Galison has noted that the militant internationalism and antitraditionalism of the Vienna Circle's manifestoes for unified science echoed the contemporary pronouncements of the Italian Futurists and the Bauhaus.[1] But the parallels are more than just rhetorical. Logical positivism was a sweeping program for the critique of culture whose basic motivation was formalist. Where geometrical form was the basis for the minimalism of modernist painting and architecture, the positivists grounded austere constraints on admissable discourse in an account of formal logic as both the structure of any meaningful language and the basis of mathematical truth. And, of course, the resulting rejection of metaphysics, religion, and traditional ethical and political discourse also belonged to the modern legacy of Enlightenment attacks on superstition and tyranny. The positivists' aim was the legitimation of those discourses which could be reconstructed in accordance with formally rational procedures. All other forms of inquiry could be rejected as noncognitive, nonsensical, or both.

1. Galison 1988: 200–201.

Although the early critics of positivism and their scientific realist and social constructivist successors have largely rejected its formalism and its opposition to metaphysics, I believe they still belong very much within the philosophical tradition of modernity, as I argued in the preceding chapter. What I take to be central to that tradition, at least in philosophy of science, is the idea that there ought to be a unified story to tell about what makes an inquiry (or its outcome) scientific (or *successful* science). The importance of such a unified story is that, like the positivists' proposed rational reconstructions, it would be of general scope and would legitimate the autonomy and cultural authority of the sciences. Richard Boyd's scientific realism is a good example.[2] Realists typically reject the positivists' attempts at a formal theory of confirmation, but they still represent science as employing a characteristic argument form, abduction. The successful employment of abductive argument serves to legitimate the authority and autonomy of the "mature" sciences, for it shows us that theories in these disciplines put us in touch with the real, mind-independent structure of the world.

Ironically, some of the most outspoken critics of both positivist and realist legitimations of science also belong within the philosophical tradition of modernity. Sociological constructivists, or radical postpositivists such as Paul Feyerabend, may deny that either positivists or realists can make good on their claims to a global legitimation of scientific knowledge, but they are in full agreement that the autonomy and cultural authority of the sciences require some such unified legitimating story. Otherwise, their insistence on the failure of the modernist projects of legitimation would not have its proclaimed cultural and political consequence of knocking the sciences down to size. Thus, I take the quintessentially modernist feature of much of philosophy and sociology of science to be the posing of a stark alternative: either realism or rational methodology, on the one hand, or relativism and the debunking of the alleged cultural hegemony of the sciences, on the other.

The debates cast in these terms have been notoriously unsatisfactory ever since they reemerged within postpositivist philosophy of science in the 1960s. But several philosophers of science later suggested a way out of this frustrating dialectic. Arthur Fine and Ian Hacking, among others,[3] cheerfully accept the postpositivist rejection

2. Boyd 1984.
3. Fine 1986a,b, 1994; Hacking 1983; Cartwright 1983; Galison 1987; Hesse 1980.

of any global legitimation of science in terms of rationality or truth but make no concessions to the allegedly relativistic or antiscientific consequences of doing so. They seem to deny that science is in any need of philosophical legitimation or that the failure of the various legitimation projects has any profound cultural or political consequences.

I want to explore further the cultural or political significance of adopting such a cheerfully postmodern debunking of the global legitimation (or delegitimation) of the sciences. Thus I am less interested in the traditionally philosophical arguments proffered in support of Fine's "natural ontological attitude" or Hacking's "experimental realism" and "dynamic nominalism" than I am in reflecting on how such attitudes redound on the cultural status generally accorded to the sciences, at least in the industrialized West. I will discuss primarily Fine's view, because he develops his opposition to any global legitimation project more centrally and explicitly, but I believe that similar arguments could be worked out for Hacking, Nancy Cartwright, Peter Galison, and others, as well as for the pronouncements on science by more familiar postmodernists such as Richard Rorty and Jean-François Lyotard.

Before proceeding, however, it may be useful to say something further about the taxonomies implicit in this project. The classification of "modern" and "postmodern" texts and practices is both slippery and controversial. Discussions of "modernity" or modernization typically invoke a disparate family of attributes—secularization, rationalization, formalism, individualism and/or the construction of the "subject," capitalism and industrialization, Western imperialism, and so forth. Quite different features often come to the fore when modernity is examined from the perspectives of politics, the arts, social theory, or science and technology. It is unclear whether "postmodernity" is supposed to represent a decisive transformation of culture, merely a new disguise for the reproduction of modernity, or a recognition that stories of modernity have always been fictions and whether postmodernity is a situation we already occupy or a possibility that still needs to be achieved (or opposed).

I focus my discussion of modernity and postmodernity in the philosophy of science around the theme of global narratives of legitimation for several reasons. This theme provides perhaps the closest thing there is to a common denominator in discussions of modernity. Emphasizing this theme also allows me the advantage of remaining uncommitted to the accuracy of the various depictions of modernity:

even those who regard the stories of modernity as fictions agree that they have been influential stories. Most important, however, the emphasis on narrative legitimation seems especially relevant to thinking about "modernity" in science and the philosophy of science. On the one hand, the development of scientific knowledge and its technological applications has been crucial to every narrative legitimation of modernity (and to the counternarratives that reconstruct the story of modern progress as one of unfolding disaster). On the other hand, a central and increasingly controversial theme throughout twentieth-century philosophy of science has been the justification for interpreting the history of science as a modernist story of progress or rational development.

We should not be deterred by the unusual bedfellows whom we find together when we think about "modernity" and "postmodernity" in this way. A philosophical taxonomy that associates Feyerabend or Harry Collins with Rudolf Carnap, Larry Laudan, and Richard Boyd but distinguishes all these from Fine, Hacking, Cartwright, and Rorty may initially seem odd. It will seem less odd if we think of "modernity" not as a position but as a shared field of conflict for which there must be a great deal of underlying agreement in order to enable sharp and consequential disagreement.[4] Realists, rationalists (in a sense that includes empiricist accounts of rational methods), and constructivists tend to agree about the significance of being able to tell a certain kind of story about the history of science. Roughly, they agree that the cultural preeminence afforded the natural sciences in the "West" is in need of global justification. It is perhaps not so difficult to imagine them all offering similar descriptions (albeit different evaluations) of the hypothetical consequences if *all* attempts at such justification (including, for diverse example, those of Carnap, Laudan, or Boyd) were known to fail utterly. By contrast, Fine, Hacking, and Cartwright, or Rorty and Richard Bernstein if one ranges more widely, are likely to view such across the board failures with some equanimity.

FINE'S NATURAL ONTOLOGICAL ATTITUDE

Fine begins his polemical papers with broad-ranging attacks on the standard realist interpretations of science and their most prominent

4. Hacking 1983: introduction.

antirealist alternatives (constructive empiricism, epistemological behaviorism, neo-Peircean pragmatism, and sociological constructivism).[5] His objections concern not just the specific positions and arguments provided by various realists and antirealists but also the shared assumptions that make the *issues* between these positions seem both intelligible and important. In their stead, Fine proposes not another position on the realist/antirealist axis but an attitude toward science. The "natural ontological attitude" is supposed to remove any felt need for a unified philosophical interpretation of science.

The path to NOA begins with Fine's observation that despite all their differences, realists and antirealists share a basic acceptance of scientific discourse. No participant in the realist debates, according to Fine, rejects the results of scientific research; each wants only to add an interpretation of what that research and its outcome really mean. Realists interpret scientists' claims about electrons as true descriptions corresponding to a definite world structure. Empiricists accept these claims as empirically adequate. Pragmatists and constructivists take them to be true in some sense that is less robust than correspondence. Fine's view is that we would do better to take scientific claims in their own terms, with no felt need to provide any further interpretation. Thus he cheerfully advises us that the naturalness referred to in NOA is "the California natural—no additives, please."[6]

Underlying Fine's California naturalism is the claim that science does for itself what the various philosophical additives were supposed to do for it, namely, situate science in an interpretive context: "What binds realism and antirealism together is this. They see science as a set of practices in need of an interpretation, and they see themselves as providing just the right interpretation. But science is not needy in this way. Its history and practice constitute a rich and meaningful setting. In that setting, questions of goals or aims or purposes occur spontaneously and *locally*."[7] In particular, Fine argues that science itself utilizes for its own purposes the supposedly philosophical concepts of truth, reality, justification, explanation, and so forth. These are not concepts that science needs but cannot provide for itself. They are the everyday material for scientific disagreement over what counts as adequate evidence, which phenomena are real rather than artifactual, and when a surprising occurrence has been satisfactorily explained.

5. Fine 1986a,b,d, 1991, forthcoming.
6. Fine 1986d: 177.
7. Fine 1986b: 147–48.

The crucial difference between philosophical accounts of these concepts and their function within ongoing scientific practice is supposed to be that in the one case, their application is global and essentialist, whereas in the other, it is local and pragmatic. Thus scientists are not concerned with whether unobservable entities exist but with whether gravitational lenses or releasing hormones exist. They would not be concerned to provide a general demonstration that the results of some "mature" sciences are approximately true; rather, they would settle for finding out whether specific results such as the Weinberg-Salam model in high-energy physics or some version of the Eldredge-Gould punctuated equilibrium view in evolutionary biology are true (or merely empirically adequate). Thus despite the claims of realists such as Boyd, abductive arguments for realism as an explanation for the success of science are *not* scientific hypotheses on Fine's view, because they occur at a different level than scientists' concerns about explanation. Scientists' concerns about explanation are localized to a particular field of investigation, and their extension beyond that field is an open question that must also be settled locally.

On Fine's account, these local questions that concern scientists can be answered quite adequately with the exercise of imagination and judgment. He illustrates what he has in mind in his criticism of van Fraassen's scruples about belief in unobservable entities.

> Are we supposed to refrain from believing in atoms, and various truths about them, because we are concerned over the possibility that what the electron microscope reveals is merely an artifact of the machine? If this is our concern, then we can address it by applying the cautious and thorough procedures and analyses involved in the use and construction of that machine, as well as the cross-checks from other detecting devices, to evaluate the artifactuality (or not) of the atomic phenomena. If we can do this satisfactorily according to tough standards, are we then still not supposed to frame beliefs about atoms, and why not now?[8]

The suggestion is that we are compelled to doubt only those scientific beliefs for which we have *specific* reasons underlying those doubts, where specific doubts are those that could suggest particular theoretical, experimental, or instrumental adjustments to assess their cogency. When all such specific doubts have been removed, or at least

8. Fine 1986b: 146.

settled according to our best judgment, there can be no further reason to suspend belief.

This emphasis on the local, scientific use of supposedly philosophical concepts does not mean that there is no place for philosophical reflection on or criticism of the ways scientific practice trades in these concepts. Nor does it rule out generalizations about their use, although Fine does suggest that such generalizations are likely to be of limited use.[9] What the adoption of NOA would require is that philosophical discussion engage the actual scientific use of these concepts and respect the contextualized concerns that circumscribe such use.

Therefore two sides exist to the natural ontological attitude and to "postmodern" philosophy of science more generally. Those I call "postmodernists" generally adopt what Fine calls a "trusting attitude" toward the sciences: "NOA trusts the overall good sense of science, and it trusts our overall good sense as well. In particular, NOA encourages us to take seriously the idea that what the scientific enterprise has to offer is actually sufficient to satisfy our philosophical needs."[10] This trust in the sciences is coupled with a thoroughgoing suspicion of interpretations of the scientific enterprise as a whole or of essential features of science which supposedly underwrite its trustworthiness: "NOA urges us to approach science without rigid attachments to philosophical schools and ideas, and without intentions for attaching science to some ready-made philosophical engine."[11]

NOA AS A TRUSTING ATTITUDE

The political issues raised by postmodern philosophy of science are highlighted by this combination of respect for the local context of scientific inquiry and resistance to any global interpretation of science which would constrain local inquiry. These issues concern the autonomy and cultural authority of the sciences. Can we distinguish a field of practices "internal" to the scientific enterprise, practices whose credibility is enhanced by their relative freedom from "external" influence and that ought to be protected from criticism or intervention on "external" grounds?

9. Fine 1986d: 174–75.
10. Fine 1986d: 177.
11. Fine 1986d: 177.

These questions are very much in the background of the various realist and antirealist interpretations of science which Fine criticized. Such views incorporate a version of the internal/external distinction and, except for some social constructivists, are intended to uphold the political autonomy and cultural authority of successful scientific practice. This concern is part of the modernist legacy of logical positivism, which had aimed to demonstrate the epistemic and cultural primacy of mathematical physics by showing that mathematics exemplified the very structure of rational thought and that sense experience was the only basis for knowledge of the world. Contemporary realists and philosophical antirealists take up somewhat different lines of defense than did the positivists: realists justify the autonomy and authority of science by showing that it gets us in touch with the real structure of the world, whereas antirealists typically show that it respects the boundaries of epistemic rationality. In both cases, however, the tacit presupposition is that it would be culturally or politically undesirable to leave the autonomy of science and the authoritativeness of its results outside the context of scientific practice itself to be the outcome of local judgment, unbuttressed by arguments that demonstrate the rationality of science in a general way (either directly or via its success in describing the real). Judgment alone, they fear, would open the way to irresolvable differences of judgment, with no way to defend scientific practice against those with fundamentally different beliefs or standards. This concern perhaps accounts for the vehemence with which many philosophers attack the views of Kuhn and Feyerabend or, more recently, of constructivist sociologists of science. Its mirror image (equally modernist) is often found in the rebellious glee with which some sociologists and philosophers claim to unmask the rationalist or realist pretensions of those who defend the authoritativeness of the sciences.

Against this background, how should we interpret the combination of Fine's trust of the sciences and suspicion toward philosophical programs for interpreting them? I want to distinguish four ways to construe this combination. These four readings exhibit a conflict between Fine's trusting attitude and his anti-interpretive stance which is focused on how one understands what is "local" in local scientific practice and how its locality is established. The implications of this conflict should incline us to endorse the fourth and most unrestrictive reading of these two aspects of postmodern philosophy of science.

The first reading takes Fine's trusting attitude toward the sciences to be fundamental, with his antiessentialism and opposition to global interpretation to follow from that trust. The resulting view would be that both criticism and defense of the general autonomy and authority of the sciences are idle because the political program of philosophical defenders of the sciences requires no defense at all but can be taken on trust. Fine's rhetoric sometimes strongly suggests this reading. He concludes his criticism of van Fraassen's constructive empiricism with "the general lesson that, in the context of science, adopting an attitude of belief has as warrant precisely that which science itself grants, nothing more but certainly nothing less."[12] He makes this point more colorfully in his response to realism: "if the scientists tell me that there really are molecules, and atoms, and psi/J particles, and who knows, maybe even quarks, then so be it, I trust them, and thus must accept [this]."[13]

But interpreting NOA as an uncritical endorsement of the autonomy and authority of the sciences seems to me a mistake. It mistakes the trust that Fine recommends for authority, and this confusion leads to serious misunderstanding. NOA does not require us to accept whatever scientists tell us, even on the rare occasions when a scientific community speaks with one clear voice. A careful reading of the texts suggests that all Fine asks is that we trust the context of concerns and practices within which particular scientific judgments are situated to be appropriate and sufficient for their assessment. Such trust does not require us to take any such judgments as authoritative. The only constraint placed by this attitude on criticism is that it be continuous with the tradition of practices and concerns constituting the field within which that judgment is situated.

Thus, at most what NOA asks us to trust are scientific traditions, where these are understood not as a consensus of authority but as a field of concerns within which both consensus and dissensus acquire a local intelligibility. The relevant notion of tradition was effectively described by Alasdair MacIntyre: "What constitutes a tradition is a conflict of interpretations of that tradition, a conflict which itself is susceptible of rival interpretations. . . . Although, therefore, any feature of any tradition, any theory, any practice, any belief can always under certain conditions be put in question, the practice of putting in question . . . itself always requires the context of a tradition."[14] So long as philosophical interpretation actually engages the pragmatically intelligible disputes that can occur within a local field of scientific

12. Fine 1986b: 147.
13. Fine 1986a: 126–27.
14. MacIntyre 1980: 62–63.

practice, NOA leaves the adequacy of that interpretation up to the exercise of reasonable judgment. This does mean that at any time one must accept most scientific practices as unquestioned in order to have resources and standards for the exercise of judgment. But which practices and results must be taken as exemplary in this way is entirely up for grabs.

NOA AND SHAPERE'S HISTORICAL INTERNALISM

So the first reading of Fine's postmodernism is untenable. But suppose we take seriously the resulting suggestion that the alternative to global interpretations of science is a reliance on particular historical traditions of scientific practice which establish a contingently coherent domain within which local, "internal" concerns can be distinguished from external and irrelevant matters. This suggests a second reading of Fine's postmodernism, one in which he resembles Shapere. By characterizing the contingent historical development of local forms of scientific rationality within particular scientific domains, Shapere tries to uphold the autonomy and authority of the sciences while avoiding anything like a global legitimation of scientific reasoning or realist success.[15] He argues that there is no essential distinction between concerns which are internal to science and those which are not but that there is a contingent distinction between those considerations which we have *learned* to regard as relevant in the course of scientific inquiry and those which have turned out to be beside the point. Thus, a strong distinction remains between internal and external issues in well-developed sciences. Shapere thinks we should respect this distinction because of its epistemic efficacy, but its formulation requires only local arguments within a particular historically constituted scientific domain.

In some ways, Shapere's position may seem quite consonant with Fine's postmodernism. Shapere presents his view as imposing no philosophical program onto the interpretation of science, takes the sciences to be historically contingent enterprises, and insists on the primacy of local arguments internal to the particular history of the domain as the only arbiter of epistemic disputes. I nevertheless believe that Shapere's defense of the internal autonomy of the developed sciences violates Fine's objections to grand narrative in that

15. Shapere 1986, 1984.

Shapere still gives a global interpretation of science as the activity of epistemic domain formation. His interpretation offers a standard narrative pattern in which to write the history of various scientific fields. This pattern provides both a criterion of scientific success and a global legitimation of the authority and autonomy of those scientific fields which meet that criterion. Shapere claims that "the very adoption of the piecemeal approach to inquiry—the laying out of boundaries of specific areas of investigation—automatically produced a standard against which theories could be assessed. Whatever else might be required of an explanation of a particular body of presumed information (domain), that explanation or theory could be successful only to the extent that it took account of the characteristics of the items of that domain."[16]

The history of scientific disciplines on Shapere's view is fundamentally a story of progress. When disciplines begin, "the motivating considerations in selecting explanatory approaches might come from just about anywhere,"[17] but they develop through a "process of gradual discovery, sharpening and organization of relevance-relations, and hence a gradual separation of the objects of its investigations and what is directly relevant from what is irrelevant thereto: a gradual demarcation, that is, of the scientific from the non-scientific."[18] His view is complicated by the ways in which consistency between domains or even their unification is scientifically valuable. Basically, however, those disciplines that contingently fail to consolidate a domain in the way he describes fall short of being scientific and have no claim to the autonomy and authority that Shapere regards as appropriate to the sciences.

The objection to Shapere's program from the standpoint of NOA is not that such histories of scientific disciplines are false; that claim would need to be examined case by case. Rather the problem is Shapere's programmatic insistence on the pattern of domain consolidation and its philosophical significance. Shapere's program is still what Fine would call "attaching science to a ready-made philosophical engine."[19] Indeed, Shapere's program promotes a characteristic feature of modernity, namely, the rationalization of autonomous domains of social practice and expertise. Even if Shapere were generally right about the actual history of scientific disciplines, we would still need a case-

16. Shapere 1986: 3.
17. Shapere 1986: 4.
18. Shapere 1986: 6.
19. Fine 1986d: 177.

by-case assessment of whether this justifies the exclusion of concerns from outside those domains. Considerations relevant to this assessment might well come from outside the domain itself. How were the boundaries of the domain constituted? What was excluded from consideration within the domain, and what were the effects of that exclusion? It will not do to insist in advance that these questions *must* be answered from a standpoint internal to the domain constituted by those exclusions.

Fine's relation to Shapere is in some ways comparable with his treatment of social constructivism. Fine rejects such striking constructivist doctrines as consensus theories of truth (along with relativism, reductionist treatments of scientific practice as "really" just discourse or social negotiation, or the insistence that scientific decision making *must* be explainable solely in terms of interests or other social factors). What is left is a constructivist research program that brackets concepts of truth and rationality and asks to what extent the activities and decisions of scientists can be explained by appeal to social factors.[20] Similarly, Fine can happily endorse Shapere's piecemeal account as an internalist research program in the history of science, the mirror image of the constructivist approach, while rejecting Shapere's aspirations toward a philosophical account of scientific rationality. But even the success of Shapere's program for any particular discipline would still leave open the question of what cultural authority thereby accrues to scientific results in its domain and even one of whether the internal autonomy that the discipline has exhibited so far ought to be sustained in the future.

NOA AS ANTIESSENTIALISM

What political consequences do these arguments have for science and its cultural context? The arguments against Shapere and metaphysical constructivism suggest that Fine will reject *any* political criticism or defense of scientific practice or belief that depends on an essentialist interpretation of scientific practice imposing a unified narrative structure on the history of science. Such interpretations supposedly aim to close off some possibilities for historical variability in what science is and how it is done, even when their essentialism is as limited as Shapere's piecemeal account of scientific domain forma-

20. Fine forthcoming.

tion. Thus, Fine insisted at one point that "the description of science as an historical entity was intended to undercut at least one version of the idea [of a science of science], the idea that science has an essence. . . . If science is an historical entity, however, then no such grand enterprise should tempt us, for its essence or nature is just its contingent, historical existence."[21] On the resulting third reading, Fine's "trusting attitude" toward the sciences would be regarded only as his assessment of what will happen when we do take seriously particularist criticisms of scientific practices. We thus seem to have come full circle from the first reading of Fine's postmodernism. There, Fine's anti-interpretive stance was derivative from and dependent on a more basic trusting stance toward local scientific practice. Now we find that Fine's trust is a highly contingent consequence of a more fundamental antiessentialism. We get the same constellation of trust and suspicion, but its significance and where it might be open to revision on the basis of sympathetic criticism have fundamentally changed.

This emphasis on antiessentialism enables Fine to tolerate some specific political criticisms of scientific practices or beliefs without abandoning his basic attitude of trust toward the sciences. He clearly believes that any inventory of what scientific practices have contributed to human flourishing which is sufficiently fairminded to appeal to our best judgment will surely call on us to preserve the substantial portion of our scientific beliefs and practices. The implicit idea that only philosophical justifications of science can preserve us from irrational rejection of science is implausible. According to Fine's view, I believe, the sciences provide what MacIntyre has called "goods internal to practices,"[22] which cannot even be appreciated without some understanding and acceptance of these practices, rather than goods defensible by appeal to general rational principles independent of historical and social context.

But this suggestion raises an interesting difficulty about how Fine should treat those interpretations of the sciences which discount or criticize the goods internal to scientific practice. His original arguments against realist and antirealist accounts of science premised that science already provided the context for its own interpretation in terms of concepts such as truth, reality, explanation, and justification. Fine cannot claim in the same way that scientific practice provides the resources for its own interpretation politically and culturally. Al-

21. Fine 1986d: 174.
22. MacIntyre 1981: 175–81.

though scientific work does require considerable attention to questions of evidence, reality, and explanation, scientists are usually professionally unconcerned with the political or cultural standing of their practices. Questions about ideology, power, social interests, gender relations, happiness, or liberation do not typically arise within scientific practice (at least in the natural sciences) in the way that concerns about truth, explanation, or reality do. When scientists are on occasion exercised to address such issues (for example, when the effectiveness of political criticisms threatens to undermine their autonomy as scientists), they address these issues not by doing more research and publishing journal articles but by stepping momentarily outside their ordinary activities and engaging in what seems to be a different sort of debate. Political issues of this sort may require something more than can be provided by the local practices and standards that could plausibly be regarded as "internal" to science.

How might Fine respond to those critics who do not share his trust in the various scientific traditions and who regard modern scientific theories and practices and their technological extensions as ideological, androcentric, antiecological, or otherwise oppressive or destructive? Do global political interpretations of the place of science in a larger social context also count as objectionable philosophical "additives"? This question is not one directly addressed by Fine, but my third reading, emphasizing his antiessentialism, suggests a definite response. If there is no essence of science, then there are also no *essential* political or cultural consequences of the authority and prominence it has within our society. It is instructive that many critics of science who do seem to adopt an essentialist line toward science or scientific rationality (for example, the Frankfurt school, some sociological constructivists, some psychoanalytically oriented feminist critics of scientific objectivity, and various other neoromantic critics of science) quite frequently accept either positivist or realist interpretations of science as defining their target. Thus the essentialist political criticisms that NOA proscribes may turn out to be parasitic on the more narrowly philosophical interpretations that were Fine's original concern.

But there may still be a conflict between Fine's trusting attitude toward science and his thoroughgoing antiessentialism. For if it is our scientific traditions that are to be taken as trustworthy, we may seem to need some account of what makes a tradition essentially scientific, or some other distinction between what is internal and what is external to the sciences. Indeed, Fine himself uses the internal/external distinction ironically to criticize philosophical interpretations of what is internal to science as themselves *external* impositions upon local scientific practice: "In science, as elsewhere, hermeneutical understanding has to be gained *from the inside*. It should not be prefabricated to meet external, philosophical specifications."[23]

Because Fine denies that we can systematically demarcate the boundary between the inside and the outside of science, we should not take his trusting attitude to depend on such a demarcation. We must instead see Fine's attitude as part of a generally trusting attitude toward local contexts of practice. No authority can accrue even to the historical tradition of scientific practice simply in virtue of its being scientific. If some of our scientific practices were to conflict with other practices or beliefs that we take seriously, NOA could counsel only the same appeal to reasoned judgment which it recommends in the case of intrascientific disputes. Once we have done our best to marshal the relevant evidence, apply locally appropriate standards, and exercise judgment to assess the outcome, there is nothing more to say; but prior to such careful exercise of local judgment, for example, on grounds of philosophical principle, we have nothing to say at all. Even with respect to the place of scientific practices within culture, then, as Fine states, "NOA, as such, has no specific ontological commitments. It has only an attitude to recommend: namely, to look and see as openly as one can what it is reasonable to believe in, and then to go with the belief and commitment that emerges."[24]

FEMINISM AND THE BOUNDS OF FINE'S ANTIESSENTIALISM

The plausibility of Fine's view on my third reading depends on the appropriateness of a twofold classification of interpretations of the sciences: they must be either essentialist, with a unified philosophical story that the interpreter brings ready-made to the sciences, or local, particularist, and without any insistence on the relevance to all sciences of any specific category (whether that category is observation or social interest, abductive argument or gender). Some recent strands of feminist criticism of the authority and legitimacy of contemporary science seem to escape this dichotomy, so they may offer a sharp

23. Fine 1986b: 148 n. 9.
24. Fine 1986d: 176.

challenge to Fine's amalgamation of trust in the sciences with an anti-essentialist suspicion of global interpretations that underwrite or undercut that trust.

Let me first separate out those strands of feminist criticism which give Fine no difficulty. He can readily endorse the relevance and appropriateness (although not necessarily the accuracy) of feminist criticisms of gender bias which appeal to and try to enlarge traditionally scientific norms of objectivity. Thus, feminist critics have frequently noted ways in which sexism has distorted inquiry, especially in the social and biological sciences, by the selective choice of experimental subjects, by the assumptions underlying the framing and answering of research questions, by the gender-blindered interpretation of results, and so forth.[25] The feminist origin of these criticisms is marked by the focus on gender, but once identified, these criticisms could be appropriated by scientific communities that do not embody feminist commitments (which is not to say that they have been or will be widely accepted by nonfeminists). Such criticisms could be acknowledged by nonfeminists as the exposure of "bad science" rather than a critique of "science as usual,"[26] even when the practices in question remain all too usual.

A postmodernist such as Fine can just as easily reject the idea that science is essentially androcentric,[27] just as he will reject Husserl's claim that science is essentially objectifying, the neoromantic claim that science is essentially reductivist, and so forth. But such a strong essentialism is more often attributed to feminists by their critics than actually asserted by feminists themselves. The most interesting cases of feminist philosophy of science, and the ones that challenge Fine's residual use of internal/external and local/global distinctions, fall in between the liberal critique of bias and the essentialist rejection of science as androcentric. Such feminist philosophers argue that gender bias is endemic to much of contemporary scientific practice, that this is not accidental but is rather deeply rooted in the development of scientific practice and its recognition as authoritative, and that reform would involve substantial changes in the ways science is practiced and/or the range of inquiries which we would recognize as scientific.

25. Representative examples of feminist criticism that focuses on sexist distortion in specific fields include Tuana 1988; Harding 1986: chap. 4; Longino 1990: chaps. 6–7; Hubbard 1990; Hrdy 1981.

26. Harding 1986.

27. This claim *might* be attributed to Griffin 1978 or Irigaray 1987.

Consider two examples that illustrate the scope of such criticisms and how they challenge the critical resources of Fine's postmodernism on this strongly antiessentialist reading. Evelyn Fox Keller argues that gendered conceptions of objectivity and power have shaped scientific practice in a variety of fields.[28] She focuses on unarticulated norms of explanation and their influence on theory construction in the contemporary interpretation and use of quantum theory, in mathematical biology and developmental biology more generally, and in genetics. Keller's arguments are not just objections to forms of unscientific bias, which could be assimilated to substantially unchanged scientific ideals, for she identifies gender at work in the most fundamental methodological concerns of particular disciplines. Her examples of gender-inflected methodological issues include: where the burden of proof lies (for example, her discussion of the pacemaker concept in slime mold aggregation); what an adequate (causal) explanation consists in and consequently what questions must be asked in scientific work (witness her reflections on "master molecule accounts" in molecular genetics, which easily extend throughout much of the biological sciences; the entire field of neuroendocrinology, for example, seems wedded to master molecule explanations); or how to assimilate significant conceptual shifts into scientific practice (as in her account of "cognitive repression" in the interpretation of quantum mechanics). If Keller's concerns are to be satisfied, substantive methodological reflection and revision would be called for in the fields she discusses.

Ruth Ginzberg's argument intersects Keller's work in an interesting way.[29] She suggests that a variety of activities that have been systematically excluded from social recognition as "science," or even as knowledge, offer an already existing model for a "gynocentric science." Where Keller looks at the gendered construction of inquiries that are widely recognized as scientific, Ginzberg looks at the role of gender in constructing the boundary between inquiries understood to be scientific and those dismissed as unsystematic and unreliable. She notes:

> In searching through women's activities outside of those that have been formally bestowed the label of "Science," I have come to suspect that gynocentric science often has been called "art," as in the *art* of midwifery, or the *art* of cooking, or the *art* of homemaking. Had these "arts" been androcentric activities, I have no doubt that

28. Keller 1985: pt. 3.
29. Ginzberg 1987.

> they would have been called, respectively obstetrical *science,* food *science,* and family *social science.* Indeed, as men have taken an interest in these subjects they have been renamed sciences and reconceived in the androcentric model of science.[30]

Ginzberg's and Keller's insistence on the far-reaching influence of the construction of gender may seem to offer a political parallel to the philosophical programs that insist on the importance to science of sense experience, abductive argument, social interests, or what have you. Even if the actual effects of gender construction on inquiry and its recognition as scientific must be examined case by case, the insistence that questions about gender are always relevant and that they must be asked even in fields of inquiry which do not seem to make gender an important category in their own terms might seem to make such criticisms a target of NOA's antipathy to "ready-made philosophical engines."

Yet neither Keller's nor Ginzberg's arguments can be easily dismissed as essentialist impositions on local scientific practice. Ginzberg's argument obviously hinges on an antiessentialist account of science, for she focuses on the cultural and political significance of how the boundaries of science and knowledge have actually been demarcated. Keller's appeal to Nancy Chodorow's psychodynamic interpretation of the construction of objectivity might initially suggest an essentialist claim that the very idea of scientific objectivity is problematically gendered.[31] But when one looks at the specific cases where she employs this analysis to look at the sciences, such an impression is not fulfilled. Keller is arguing that the working out of scientific conceptions of knowledge, explanation, and objectivity cannot be isolated from more general cultural and cognitive patterns. This dependence has nothing to do with intrinsic features of science or of objective inquiry and everything to do with the larger cultural context within which particular scientific disciplines have actually developed. Because of the pervasive gendering of cognitive activity in the dominant culture, gender can function constitutively within those scientific concepts and practices without having been identified as such explicitly. Yet this does not guarantee that any or all scientific practices will be debilitatingly gendered; the presence and actual impact of gender in science must be demonstrated case by case. Furthermore, and per-

30. Ginzberg 1987: 91–92.
31. Keller 1985: pt. 2.

haps most important, the concept of 'gender' itself is not unproblematic in Keller's work, as her treatment of Barbara McClintock's research shows. Keller's reflections on science and gender leave neither science nor gender unchanged.

Furthermore, if Fine were to make this argument, thereby reaffirming a stronger version of his trusting attitude toward the sciences specifically, he would undercut his antiessentialism by privileging a particular account of what is internal to the tradition of the sciences. The objection to Ginzberg or Keller would be that their focus on gender was a large-scale program imposed on science from the "outside"; but their argument is that gender functions very much "inside" the sciences, precisely by helping to shape what counts as inside or outside. Shapere, who provided the model for our second reading of NOA's postmodernism, bases a philosophical program on taking at face value scientists' own interpretation of how their domains of inquiry are bounded. Fine, however, must reject both that critical hermeneuticism which insists that this face value must mask a deeper truth, and Shapere's, which denies that the constitution of scientific domains is open to suspicion. To do otherwise is to beg the question between an internalist such as Shapere and the feminist critics.

This recognition offers a fourth reading of Fine's postmodernism, one that I think Fine's arguments should ultimately lead us to endorse. On this reading, what is at issue in reflecting about science is how a particular domain of concerns comes to count as a coherent field of inquiry. Because of what is at issue, Fine rightly requires that "NOA does not prejudge the constitution of the scientific world; that is, whether the scientific facts and objects are essentially social or essentially objective, or whatever. Its attitude is to let the chips fall where they may."[32] From the standpoint of the philosophical tradition of reflection on science, this may seem to place the philosophy of science more within the realm of social and political inquiry rather than its traditional home within epistemology and metaphysics. But that would be to miss the antireductivist force of this reading of Fine. It is as much a mistake to reduce science to politics or social construction as it is to restrict it to epistemology. Political, epistemological, sociological, or psychological accounts are on a par at the outset when we ask how (and whether) a distinct domain of inquiry has been successfully or appropriately constituted. There is no line of argument which can reassure us that we have (and will continue to have)

32. Fine forthcoming.

the right language and the right locus of concerns to understand science.

This fourth reading of Fine's postmodernism reduces the conflicts we have found between trust in local practice and suspicion of global readings by glossing both views in very unrestrictive ways. In contrast to its treatment by the third reading, NOA's antiessentialism would still leave considerable room for large-scale social criticism of the modern sciences (and for responses and countercriticisms of comparable scale), so long as they situate themselves within particular historical contexts and the critical resources specific to those contexts.[33] NOA's trusting attitude would remain, but only in the sense that Fine trusts that the contingencies of history give us all the resources necessary to tell us what we need to know about science. This says nothing yet about which contingencies these are.

Thus, on this fourth reading of NOA, Fine's frequent reference to trust in *science* is either misleading or mistaken, insofar as "science" is identified with any determinate interpretation of what does or does not count as scientific. Asking that we take science "on its own terms" should not invoke an authoritative account of what those terms are. "Its own terms" denotes a field of interpretive dispute rather than a definitive vocabulary. Thus, if Ginzberg is right, understanding science on its own terms requires us to understand the ways in which the authentication of knowledge is gendered, leading to exploration of gossip, midwifery, and cooking, and the history of sexism, along with physics and geology. But there are no philosophical shortcuts. The only way to find out whether she is right is to look where she is pointing and see whether we can come to see what she finds there. Likewise with realists, constructivists, empiricists, and neoromantic visionaries.

This account of the critical resources available to someone who adopts Fine's recommended attitude may seem anemic in its lack of general criteria on which to ground one's judgment and its openness to unfamiliar interpretations of science. Fine himself recognizes that his appeal to good judgment is empty without the detailed work necessary to provide a local and particularist grounding for judgment in any particular case. This recognition is evident in his recommendation that

33. A similar position has been defended by Fraser and Nicholson (1988) against what they take to be the apolitical postmodernism of Lyotard and Rorty; my argument is deeply indebted to Fraser and Nicholson.

> an open, social particularism . . . is the right corrective to philosophical (especially realist) distortions of science, and the place where lots of good work can be done. Among the work to be done is to achieve some understanding of what is actually involved in rational acceptance and proof in science, of what, in Boyle's words, deserves "a wise man's acquiescence." This job involves exploring the diverse range of contexts, historical and contemporary, in which inquiry is carried out.[34]

Such a recommendation would still be vacuous if taken out of context. What contemporary philosopher of science would deny having paid attention to "the diverse range of contexts in which inquiry is carried out?" But Fine has provided detailed arguments and strategies for extending them which show how many treatments of science which initially seemed historically sensitive and particularist could instead be cases of "attaching science to a ready-made philosophical engine." The criteria for "wise acquiescence" can never be fixed once and for all, for they are inescapably open-ended and context-sensitive.

In the case of feminist philosophy of science, we can perhaps see more clearly how Fine's program is less anemic than it initially looks. Keller's and Ginzberg's arguments gain credibility from several decades of extensive feminist research in a variety of fields, research that has repeatedly exposed and challenged the gendering of scholarly inquiry and authenticated knowledge. Such research would not have succeeded without frequently and substantially satisfying previously established scholarly standards, but it also partially transformed those standards in so doing. Without that range and depth of detailed research as background, Keller's and Ginzberg's claims would undoubtedly be unconvincing. But in that context, the burden of proof has to some extent now been shifted.[35] It is now more credible that the substantive ideals of explanation in physics or biology are not gender-neutral; it is also now reasonable to demand a strong and detailed argument before expecting that the natural sciences will be different from other fields in being relatively uninflected by gender. These considerations alone do not suffice to make Keller's or Ginzberg's case, for we must still evaluate their specific arguments and evidence in this context. These considerations do, however, illustrate how the appeal to reasonable judgment in specific historical contexts

34. Fine forthcoming.
35. Fuller 1988: chap. 4.

is not entirely unconstrained and how such standards of judgment might still be flexible and open to change. Just as it has always been reasonable to accept, reject, or suspend judgment on various knowledge claims despite the absence of any plausible general theory of confirmation, it is also reasonable to accept, pursue, or reject various approaches to a political understanding of various scientific practices and disciplines in the absence of a general philosophical interpretation of science and its aims.

Such a historically sensitive and open-ended particularism is what I take to be fundamental to the natural ontological attitude on this fourth and final reading. When we adopt such an attitude toward science and its epistemic and political interpretation, we relieve the specific and fundamental conflicts disclosed within the other readings of Fine's postmodern philosophy of science. More important, we take up an approach to the sciences which avoids both science-bashing and the foreclosure of the intellectual resources needed to raise and begin to answer critical questions about the cultural and political significance of the sciences. Much more remains to be said about the positive implications of such an approach to the philosophy of science and how it would relate to other discussions of "modernity" and "postmodernity." But the difficulties with alternative readings of Fine's NOA suggest that such a thoroughgoing postmodernism offers the best hope for what Fine once called "a decent philosophy for postrealist times."[36]

36. Fine 1986a: 113.

[3]

Arguing for the Natural Ontological Attitude

In this chapter I am concerned with Fine's reasons for adopting the "natural ontological attitude" rather than his arguments against realism and antirealism. Unless these reasons are adequate, Fine's papers are likely to set off a scramble for new and better arguments concerning realism rather than the reorientation of philosophy of science which his polemics are clearly intended to provoke. Moreover, the significance of this discussion goes beyond the particulars of Fine's proposal; the underlying issue concerns what kinds of arguments are needed to justify dispensing with the legitimation project. Thus, for the purposes of this chapter, we can presume that Fine's criticisms of other positions are generally sound. Our question is whether his proposed alternative attitude can then satisfy the same argumentative standards.

Before we can examine the arguments Fine provides for NOA, we need to note how he argues against various realist and antirealist positions. This summary will enable us to consider what NOA is and to what positions this attitude commits us. We can then proceed to consider what arguments Fine can offer for NOA, beyond his criticism of alternative positions.

Although he has other criticisms, Fine's principal objection to the standard explanationist defenses of scientific realism is that they beg the question against the antirealist. As Fine construes the debates between realists and instrumentalists, the issue is whether any of the epistemic virtues (for example, explanation, coherence, simplicity) are truth-conferring. Realists typically believe that our best explanations, or perhaps our most coherent theories, are true, whereas instrumentalists counter that these virtues may give us reasons to use theories but not to believe them. Because of this disagreement, the realist cannot appeal to arguments to the best explanation or arguments from coherence to justify the truth of realism without assuming what is at issue, namely, the truth-conferring power of such arguments.[1] Fine illustrates this with an analogy to metamathematics. A set-theoretic proof of the consistency of set theory offers no rational reason to accept set theory. Likewise, a realist abduction to realism should not convince an open-minded inquirer. Fine argues that the various antirealist views fare no better. Such views either provide an alternative semantics to a realist interpretation of truth (as correspondence to a definite, mind-independent world structure) or they argue on epistemological grounds that something less than full-blooded belief is an appropriate epistemic attitude toward some or all scientific theories. Fine claims that the views of the "truthmongers," as he calls the semantic antirealists, are deeply incoherent because they must utilize an unanalyzed notion of truth to render their analysis of truth (for example, as idealized rational acceptability) intelligible. Epistemological versions of antirealism (such as empiricism) turn out on Fine's view to be either arbitrary impositions on scientific practice or themselves unintelligible. They are arbitrary if they use a philosophically specified distinction between the observable and the unobservable to determine the limits of reasonable belief in the results of science. But if we avoid the charge of arbitrariness by allowing science itself to determine what is observable, as van Fraassen does,[2] then the view becomes unintelligibly circular. We cannot know which parts of science to believe (rather than merely accept as empirically adequate) until we know which phenomena are observable. We cannot know that, however, unless we have reason to believe those parts of science which tell us what is observable.

Fine's conclusion from these various arguments is that no rational reason exists to accept either the inflationary metaphysics of realism

1. Note that the realist cannot escape this dilemma by concluding that realism is at least empirically adequate. To say that scientific realism (as a theory purporting to explain the contingent and empirically discovered success of science) is empirically adequate is simply to deny realism. Realism would be empirically adequate if our scientific theories were empirically adequate. But the latter claim is just an empiricist version of antirealism.

2. Van Fraassen 1980.

or the inflation of epistemology or semantics which is necessary to substitute an antirealist alternative. His deflationary alternative (NOA) seems to incorporate four central points:

(1) Science is a historically contingent enterprise, with no essential characteristics

(2) Scientific practice typically makes extensive local use of supposedly philosophical concepts such as truth, reality, justification, and explanation, uses that are governed by the practical needs of scientists within their social and historical situation

(3) The exercise of imagination and especially judgment within the local scientific context is sufficient to settle most issues that arise within the research context (which still leaves considerable room for reasonable disagreement about specific scientific issues on local grounds)

(4) No justification exists for overriding local scientific uses of these philosophical concepts on the basis of global philosophical interpretations of their application to science, hence philosophers should quit looking for the kind of global view of science exemplified by realism or antirealism

We can now ask what arguments Fine offers in defense of these claims and whether his arguments are adequate. Clearly, much of their support is supposed to derive from the specific arguments Fine has offered against various realisms and antirealisms. But the need for additional argument should be evident from the programmatic character of NOA. Fine does not merely reject the specific realism and antirealist positions in the literature. One could conceivably reject all the available philosophical positions but still insist that it is important to find more-adequate realist or antirealist alternatives. To adopt NOA is to suggest instead that we stop asking the questions to which these positions were answers and refuse to look for or accept any other philosophical interpretations of science which would purport to be answers to those questions.

Fine himself oscillates uneasily between two distinct strategies of argument for NOA, and both these strategies encounter prima facie difficulties of the same sort he attributes to realism and antirealism. In the end, I shall argue, one of these strategies offers considerably greater promise for extricating itself from these difficulties than does the other, but much more work needs to be done to redeem this promise.

Fine most frequently appears to recommend a minimalist strategy that initially seems both appealing and highly plausible. The minimalist strategy emphasizes that NOA is an attitude rather than an alternative philosophical position, suggesting the appropriateness of a more modest sort of argument. On this approach, the arguments against realism and antirealism motivate NOA more than they argue for it. The strategy has two stages. The first stage reiterates the arguments that the various attempts so far to place science within an ontological, epistemological, or semantic context have turned out to be arbitrary, unintelligible, or question begging. The second stage notes the lack of any terrible consequences to doing without such interpretations (after all, if the first arguments are right, science has managed quite well without any good philosophical interpretations) and asks us, Why not try NOA and see if we can learn to like it? In using this strategy, NOA appeals simply to one's judgment as to whether there is any point to continuing to look for a general interpretive framework for science. Against the philosopher who nevertheless insists on the importance of such a philosophical interpretation of science, NOA has no further argument to offer; it can ask only that she reconsider whether such a project really has a point any more. NOA thus treats philosophy in the same way it treats science, claiming on the basis of local, historically situated philosophical judgment that inflationary metaphysics, epistemology, or semantics no longer seem to be worth pursuing.

This interpretation of NOA has the advantage of a kind of consistency and it is explicitly supported in Fine's text,[3] but it also looks suspiciously prone to the sort of circularity for which Fine criticized explanationist defenses of realism. Fine is appealing to our local, contextual judgment to argue that only local, contextual judgment is adequate for the assessment of claims about truth, justification, and explanation. But the general issue between Fine and both the realists and the antirealists is whether our local, contextual judgment, unbuttressed by further philosophical argument, can give us an adequate understanding of science and the adequacy of its claims. Realists and antirealists demand a metaphysical, epistemological, or semantic grounding for the judgment that one should "accept confirmed scientific theories in the same way we accept the evidence of our senses"[4] because they believe that such judgment alone is untrustworthy, thus

3. Fine 1986b: 141 n. 4.
4. Fine 1986a: 127.

opening the way for an unacceptable relativism or irrationalism. To respond by appealing to the same sort of judgment at a metatheoretical level is to commit the same fallacy that Fine ascribes to the realist, who presupposes the truth-conferring power of arguments to the best explanation in science.

I do not see any clear way out of this difficulty if such strategy of argument is what Fine is proposing for NOA. It will not help to say that NOA is only an attitude (and therefore presumably does not require the same sort of argument that a philosophical position would). If it is a reasoned attitude, then the question at stake between its proponents and realists or antirealists should not be begged. If it is not a reasoned attitude, Fine confirms the worst suspicions of his opponents. Nor will it help to say that this sort of metatheoretical circularity need not be forsworn; such acceptance would rehabilitate realism along with NOA, for a parallel argument provided Fine's principal objection to realism. However appealing it may have seemed initially, the minimalist strategy for motivating NOA fails.

There is an additional difficulty. The success of this strategy depends on whether NOA satisfies the best judgment of those sufficiently knowledgeable to make reasoned judgments about it. Fine claims that NOA offers a better philosophical approach to science than do its realist and antirealist opponents. But it seems clear that the reasoned judgment of his philosophical peers is not generally in accord: realism or antirealism still seems to command the allegiance of most philosophers of science. Perhaps more widespread consideration of Fine's papers will change this situation. But unless that happens, NOA's appeal to philosophical judgment seems not to redound to its favor.

Fortunately, the language of the NOA papers sometimes suggests that a less modest and perhaps more promising strategy of argument is being proposed. The "metatheorems" proposed in "Unnatural Attitudes"[5] suggest not an appeal to judgment that the various realisms and antirealisms are unpromising but an argument in principle that such interpretations *cannot* work. Such an intention also seems suggested when Fine says that "NOA tolerates all the differences of opinion that science tolerates, [but] does not tolerate the prescriptions of empiricism or other doctrines that externally limit the commitments of science."[6] If the rejection of philosophical interpretations of sci-

5. Fine 1986d.
6. Fine 1986b: 150.

ence were the outcome only of considered judgment, we might expect NOA adherents to be more tolerant of dissenting judgments.

Fine also gives us an indication of where to look for the sorts of arguments which would justify such intolerance. Its justification is found in antiessentialist arguments about truth, justification, explanation, and the like. As Fine says:

> The antiessentialist aspect of NOA is intended to be very comprehensive, applying to all the concepts used in science, even the concept of truth. Thus NOA is inclined to reject *all* interpretations, theories, construals, pictures, etc., of truth, just as it rejects the special correspondence theory of realism and the acceptance pictures of the truth mongering antirealisms. For the concept of truth is the fundamental semantic concept. Its uses, history, logic, and grammar are sufficiently definite to be partially catalogued, at least for a time. But it cannot be "explained" or given an "account of" without circularity.[7]

Fine's argument for this claim is found in his various attempts to show how such circularity arises in specific attempts to give an account of truth[8] and in his suggestion that the import of these arguments is quite general. The basic strategy of Fine's arguments is this: to fix the nature of truth (for example, as idealized rational acceptance), one must presuppose an unanalyzed conception of truth to be applied within the analysis of truth itself. Typically, he argues, this circularity prevents us from ever actually seeing what the analysis of truth tells us. In accounts of truth as idealized rational acceptance, for example, "it looks like what we are called upon to do is to extrapolate from what is the case with regard to actual acceptance behavior to what *would be* the case under the right conditions. But how are we ever to establish what is the case, in order to get this extrapolation going, when that determination itself calls for a prior, successful round of extrapolation? It appears that acceptance locks us into a repeating pattern that involves an endless regress."[9] Such an analysis of truth is self-defeating, because if truth needs an analysis, we can never determine what the analysis is.

In making these arguments, Fine accepts that the concept of truth is contextually situated, stating that "the concept of truth is open-

7. Fine 1986b: 149.
8. Fine 1986b: 139–42, 1986d: 169–70.
9. Fine 1986b: 141.

ended, growing with the growth of science; . . . the significance of the answers to [our questions about truth] is rooted in the practices and logic of truth judging,"[10] but he rejects any *theory* that construes truth as contextual. Although Fine does not explicitly develop these arguments, NOA also seems to call for similar treatments of justification (as holistic and pragmatic), explanation (as contextual and pragmatic), and existence (as not a real predicate). Fine should be no more happy with a contextualist theory of justification which attempted to show how a sentence in question must be related to other sentences and practices to count as "justified" than he would be with an acceptance theory of truth. Yet what NOA stands for above all is a pragmatic turn to local contexts of use to settle any questions that arise concerning supposedly philosophical concepts such as truth, existence, justification, or explanation.

The question that Fine's arguments insistently raise is why his own discussions of these philosophical concepts are not themselves the sort of global philosophical interpretations prohibited by NOA. His specific arguments against realism and antirealism repeatedly aim to expose self-defeating circularities by turning their accounts of truth, justification, and explanation against themselves. If NOA is to be an adequate replacement (or displacement) of such views, it should itself be free of this sort of circularity. Yet it is evident where such circularity is likely to manifest itself. NOA develops a distinction between global and local interpretations of scientific practices to show (metatheoretically) that global criticisms or proposed constraints that do not engage specific scientific concerns locally are arbitrary, unintelligible, or question begging. The question that naturally arises, however, is whether this distinction between global and local criticism is arbitrary, unintelligible or question begging.

The problem is that Fine's application of the distinction seems to be based on global interpretations of fundamental philosophical concepts of precisely the sort that NOA aims to prohibit. And, ironically, his "no-theory" of truth (and ceteris paribus, similar discussions of justification, explanation, and existence) does for philosophy what NOA says philosophy cannot do for science: it rules out globally and in advance certain kinds of positions and arguments and does so on the basis of a general conception of what a philosophical account or explanation is. Why aren't all these claims underlying NOA just more philosophical interpretations which have no local justification within

10. Fine 1986b: 149.

scientific or philosophical practice and which consequently require NOA to be turned against itself?

It is not hard to see why Fine thinks that his discussions of truth, reality, justification, and explanation are different from realist or antirealist accounts and why he calls them "no-theories."[11] The realist demands that scientific theories must have converging ontologies if their terms are to count as genuinely referential; the empiricist limits justification for belief strictly within the bounds of what is observable; the neo-Peircean truthmonger stipulates that we should accept only those beliefs around which an ideally rational consensus might form; and so forth. NOA's philosophical concomitants place no constraints of this sort on the practice of science, and on Fine's view, this restraint is its principal virtue as well as its distinction from other interpretations of philosophical concepts. The question that he needs to answer if NOA is to be viable is why this distinction between global views that constrain local practice and those which do not is so important, and in particular, how this distinction rescues NOA from the charges of arbitrariness, unintelligibility, or question begging which he believes plague the philosophical views he rejects.

I can see only three possible responses to this question which would be prima facie plausible. First, Fine could claim that his own global discussions of truth, justification, and so forth are of heuristic and/or polemical value only; NOA does not depend on the acceptance of these views (this approach, however, revives the first strategy of argument for NOA, which I have already shown to fail). Second, he could claim that his own apparently global discussions of truth, justification, and explanation are not really philosophical accounts of their "nature" but something less than (and also less objectionable than) a philosophical theory, perhaps in a way suggestive of Rorty's distinction between systematic and edifying philosophical projects.[12] And third, Fine could accept that his own discussions are philosophical and global in the same way as realism and antirealism but are preferable because they offer better philosophical accounts of these concepts: what NOA objects to on this reading would not be philosophical interpretation, per se, only bad philosophy.

The prospects for these responses seem to me quite uneven. Having already argued that the first response faces insuperable problems, I can move directly to the suggestion that Fine is not really offering

11. Fine 1986d: 176–77.
12. Rorty 1979: chap. 8.

philosophical accounts of truth, justification, and so on. This second response also has two difficulties that must be overcome if it is to be promising. First, Fine's views really do seem like philosophical positions. His view that "truth is the fundamental semantical concept"[13] that all intelligible discourse presupposes or his implicit view of justification and explanation as inevitably contextual and pragmatic seem straightforwardly philosophical. They impute a unity to our use of these concepts which derives from their function rather than their content, but this does not seem to make the sort of difference that would challenge their status as philosophical or global. Second, the difficulty would seem to be displaced rather than removed by this ploy. If Fine's own views are to escape NOA's strictures against global philosophical interpretations because his views are not really "philosophical," then we need an account of what justifies this appeal to the boundaries of "philosophy." If NOA were to be antiessentialist about everything philosophical except "philosophy" (which is taken to be essentially essentialist), the reasoned acceptance of NOA would still depend on the kind of global philosophical view that Fine wants to eschew. This point illustrates the intractability here of the more general problem so elegantly displayed in Fine's own arguments against realism and antirealism. Fine uses a metatheoretical distinction between global and local uses of philosophical concepts to escape the reflexive application of his own thoroughgoing antiessentialism. But this distinction seems vulnerable in any of its possible guises to similar imputations of unintelligibility, arbitrariness, or question begging. Just as Fine doubts that more sophisticated versions of realism or antirealism are worth pursuing any more, it is not clear how a more sophisticated, antiphilosophical NOA would fare any better either.

These considerations lead us to the third response. Here, the argument strategy at work in Fine's criticisms of truth-mongering antirealisms would seem an appropriate model for NOA as a thoroughly philosophical antiessentialism. Fine sketches arguments that begin from the fact that "there is no form of life, however stripped down, which does not trade in truth" and that "the redundancy property of truth makes truth a part of any discourse that merits the name."[14] To have the general import that this interpretation of NOA would require, these arguments would then need to be developed to show that such use of an unanalyzed notion of truth, which places no constraints

13. Fine 1986b: 149.
14. Fine 1986d: 170.

on what can be true, is a necessary condition for "any discourse that merits the name." If this could be shown and if similar arguments were developed for the necessity of unanalyzed, contextualized *uses* of (rather than interpretations of) concepts such as justification, explanation, and existence within the practice of science, then NOA would not be vulnerable to charges of reflexive circularity.

These arguments have parallels in Wittgenstein, Heidegger, and Davidson.[15] More recently, Mark Okrent has argued that Richard Rorty's pragmatism also requires such arguments, for similar reasons, to justify its own philosophical underpinnings.[16] Along with Okrent, I believe these are transcendental arguments. In their most familiar form, such arguments display necessary conditions for the possibility of a given capacity (for example, experience of objects, intentionality, or language), which in turn show the ontological limitations of what can be manifest through that capacity. Kant's Highest Principle of Synthetic Judgment claims, for example, that the necessary conditions of any possible experience are the ground for synthetic a priori knowledge about objects of experience, which must be spatiotemporal, causally connected, and so forth.[17] Okrent rightly argues that the transcendental arguments needed by pragmatists have no analogue to Kant's Highest Principle, for they show the necessarily contextual and holistic character of meaning, justification, and truth but place no constraints on what could be meant, justified, or true. These arguments also place no a priori constraints on what could compose the contexts that manifest meaning, justification, and truth.

In this chapter I strongly suggest that such contextualist transcendental arguments are needed to justify dispensing with the legitimation project. In Part II I work out such strongly contextualist understandings of meaning, justification, and truth within the specific context of the philosophy of science. My account makes no serious attempt to show that these are necessary conditions for intentionality

15. Wittgenstein 1953, Heidegger 1977, Davidson 1984.

16. Okrent (1989) characterizes these arguments for pragmatic antiessentialist accounts of meaning, truth, and justification as transcendental arguments that attempt to display the necessary conditions for the possibility of a discourse having semantic content. They differ from classically Kantian transcendental arguments in rejecting the Highest Principle of Synthetic Judgment, according to which the transcendental conditions for the possibility of experience are also constraints on what can be an *object* of experience. This rejection of the second stage of Kantian transcendental argument aptly suits Fine's discussion of NOA, for Fine similarly wants to claim that the fundamentally philosophical concepts that are constitutive of local scientific practice and discourse place no constraints on what can count as a real object, as genuine evidence or explanation, or as truth.

17. Kant 1965.

or knowledge, which would require a very different book and much more extended argument. My more limited aim has been to articulate the philosophical context within which Fine's and my arguments against the legitimation project could be compelling and to suggest its implications for understanding and assessing critically the practices and achievements of the sciences.

[4]

Should Philosophy of Science Be Postpositivist?

"Once upon a time, we were all logical positivists. . . ." So begin the stories that we philosophers of science often tell ourselves in making sense of the current preoccupations and concerns of our discipline. Our stories tell of the collapse of positivism and the resulting fragmentation of the discipline's nostalgically recollected unity. As is our custom, we tell of positivism's decline in terms of discoveries and reasoned inferences. Around 1960, philosophers of science rediscovered that science had a history and recognized that the logical positivists' idealizations were ill suited to the complexities of the historical record. At the same time, philosophers began to attribute epistemic primacy to scientific theory, overcoming empiricist scruples about the meaning of theoretical terms and the scope of physical laws. Observation and methodology turned out to be already laden with the interconnected theoretical commitments that constituted scientific research programs, paradigms, or domains of inquiry. Scientific communities also came to philosophical prominence, as theoretical research programs were increasingly ascribed to scientific communities rather than to individuals. Metaphysics regained its respectability among philosophers of science: causes reemerged from behind regularities, and scientific realism became not just a live option but perhaps the most widespread philosophical position as well. Naturalism became the preferred metaphilosophy, as logical analysis of meanings was no longer seen to provide philosophy of science with a domain or a method free from empirical contamination.

As befits a supposedly fragmented discipline, our stories have different endings; some culminate in realism, others in historicized rationalisms, cognitivist naturalisms, and disunifying turns toward studies of the special sciences. Many of our stories about the postpositivist rediscovery of history, theory, community, and reality also serve as cautionary tales. They warn us of the dangers of going too far in any of these directions: a philosophical history of science must still be internal history; the theory-dominance of postpositivist philosophy must not neglect experiment; recognition of the importance of scientific communities must not slip toward social constructivism and epistemic relativism; perhaps only a "moderate" or restricted realism will do. Meanwhile, naturalism should promote a philosophy of science which is continuous with the empirical disciplines without being superseded by them.

Yet these stories of the philosophical turn away from positivism are increasingly at odds with some of the most interesting work in science studies. At the most straightforward level, historical reconstructions of the positivist tradition suggest that our familiar folk renditions of that history border on caricature.[1] But, as I argue in this chapter, two other conflicts could have especially significant consequences for the future direction of philosophy of science. First, some of the best contemporary philosophical studies of science feature compelling links with positivism. The work of Nancy Cartwright, Ian Hacking, Arthur Fine, and Bruno Latour in particular might be better understood as extending the positivist tradition rather than as attempting to supersede it. Second, more subtly, the disciplinary self-conception of philosophy of science as postpositivist turns out to be an impediment to a constructive engagement with both feminist studies of gender and science and emerging historical and anthropological work influenced by interdisciplinary cultural studies. This chapter is an exploration of these two cracks in our postpositivist disciplinary facade.

Understanding the continuity between logical positivism and more recent philosophical studies of science requires some reconception of the positivist legacy. The Anglo-American assimilation of Central European positivism has emphasized the union of a renewed Humean empiricism with Russell's and Wittgenstein's exemplary reconstructions of logical form. Yet underlying the positivists' use of formal logic and their somewhat grudging slide toward empiricism was their understanding of the sciences as offering a powerful antidote to su-

1. Friedman 1984, 1987; Coffa 1991; Richardson 1990; Ryckman 1991a,b; and others.

perstitious and mystifying ideologies. The sciences challenged such mystifications by insisting that discursive claims be restricted to matters that could be made publicly manifest. The positivist opposition to "metaphysics" was directed against reified sources of cultural and political authority which could not be checked by publicly accessible means. Some recent work in philosophy of science can thus be understood as more radically positivist in spirit in its exposure of psychological empiricism and semantic realism as themselves unnecessary reifications in need of positivist critique.

A more radically positivist critique of philosophical empiricism is readily apparent in the work of Cartwright, Hacking, and Latour. All three philosophers interpret scientific practices and claims in terms of what can be made publicly manifest, but they reject empiricist conceptions of sensation or observation as grossly inadequate or irrelevant interpretations of manifestation. Instead, they emphasize practices of manipulation, measurement, and marking which are registered by instruments rather than sense organs. Cartwright's work on causes and capacities might initially seem to be strongly antipositivist, for she defends the existence of capacities and singular causes against familiar empiricist scruples.[2] Yet she argues instead for a reformed, antiempiricist positivism.[3] Against the traditional empiricist picture of science as founded on sensory observations, Cartwright insists that "science is measurement." Measuring is always a publicly accessible practice in which scientists interact with things rather than register their sensory effects. Capacities, in turn, are what can be measured, and they must be accepted as an ineliminable part of scientific practice. But instead of abandoning positivist qualms about unnecessary and misleading reification, Cartwright redirects them against the Humean acceptance of regularities and laws: "The pure empiricist should be no more happy with laws than with capacities, and laws are a poor stopping-point. It is hard to find them in nature and we are always having to make excuses for them. . . . [My] point is to argue that we must admit capacities, and my hope is that once we have them we can do away with laws. Capacities will do more for us at a smaller metaphysical price."[4]

2. Cartwright 1989.

3. Cartwright herself tries to preserve the term 'empiricism' for her project (as I once did in Rouse 1987b in referring to her as a prototypical "new empiricist"), but for present purposes, it is useful to restrict that term for commitments to the primacy of sense experience in order to clarify both Cartwright's continuities and her divergences from the positivist tradition.

4. Cartwright 1989: 8.

Hacking's tendency toward an antiempiricist positivism is more modulated. He emphasizes the multiplicity of scientific practices and aims and encourages a tolerant skepticism toward almost all general pronouncements about the sciences. Yet in contrast to empiricist emphases on observation ("observation is over-rated"[5]), Hacking is adamant about the central place of experimentation and intervention in scientific practice. Moreover, the manipulation of things to produce reliable effects holds a fundamental place in Hacking's overall picture of the sciences, for all their diversity. Thus, he notes that "the best kinds of evidence for the reality of a postulated or inferred entity is that we can begin to measure it or otherwise understand its causal powers. The best evidence, in turn, that we have this kind of understanding is that we can set out, from scratch, to build machines that will work fairly reliably, taking advantage of this or that causal nexus."[6] Hacking is too cautious and too tolerant to substitute engineering for observation in a radical positivism that would have us avoid all speculation about "unmanipulables." He nevertheless makes his own inclinations quite clear. Confessing "a certain distaste for occult powers," he offers "a slender induction: long-lived theoretical entities, which don't end up being manipulated, commonly turn out to have been wonderful mistakes."[7]

Latour may seem quite out of place in a listing of contemporary fellow travelers with positivism. Yet his account of how objects and facts are constituted and sustained through practices of network building is strongly akin to operationalism. Latour denies, for example, that time and space are unified frameworks within which actions are already situated. Instead, they are locally constructed frameworks that depend on specific practices of measuring, comparing, and mapping. Thus, he claims that "we do not live *inside* space, that has billions of galaxies in it; on the contrary, this space is generated *inside* the observatory by having, for instance, a computer count little dots on a photographic plate."[8] Similarly, Latour states:

5. Hacking 1983: 168; Fine clearly shares this inclination to deny any special epistemic significance to observation. He responds to van Fraassen's resurrected empiricism by suggesting that "we can make the question of observability, as a warrant for belief, very acute by asking, why restrict the realm of belief to what is observable, as opposed, say, to what is detectable? I think the question is acute, because I cannot imagine any answer that would be compelling" (1986b: 146).

6. Hacking 1983: 274.

7. Hacking 1983: 275.

8. Latour 1987: 229.

> Time is not universal; every day it is made slightly more so by the extension of an international network that ties together, through visible and tangible linkages, each of all the reference clocks of the world and then organizes secondary and tertiary chains of references all the way to this rather imprecise watch I have on my wrist. There is a continuous trail of readings, checklists, paper forms, telephone lines, that tie all the clocks together.[9]

Not just one space and one time exist, but many, and they are commensurable only to the extent that their commensuration is actively achieved and constantly policed. The networks that incorporate instruments, inscriptions, and the practices of using, transporting, and coordinating them perform such commensurations, and thereby enable anything to manifest itself in a consistent and reliable way and thus to count as (the same) fact, object, performance, or setting. Indeed, Latour denies that there can be "a fact, a theory or a machine that could survive *outside* of the networks that gave birth to them."[10] Once again, we have both a reluctance to countenance the existence of things that are not manifest and an account of manifestation which emphasizes public practices of manipulating, transporting, inscribing, and juxtaposing, rather than perceptual experiences. Indeed, Latour objects not only to the empiricist reification of sensation but also to any explanatory appeal to cognitive processes more generally. In place of discussions of "the mind and its cognitive abilities," he insists that

> any study of mathematics, calculations, theories and forms in general should do quite the contrary: first look at how the observers move in space and time, how the mobility, stability and combinability of inscriptions are enhanced, how the networks are extended, how all the informations are tied together in a cascade of re-representation, and if, by some extraordinary chance, there is something still unaccounted for, then, and only then, look for special cognitive abilities.[11]

Latour exudes confidence that categories such as perception, cognition, and rationality would thereby show their dispensability for any understanding of science or the world.

9. Latour 1987: 251.
10. Latour 1987: 248.
11. Latour 1987: 246–47.

Latour's hostility toward any account of scientific cognition would probably not be widely shared among the other philosophers whose work I would assimilate to an extended positivist tradition, although Hacking does explicitly dismiss explanatory appeals to scientific rationality.[12] Yet much greater continuity follows if one understands Latour's hostility to 'cognition' as part of a broader criticism of realism about *meaning*. I think Latour objects to "cognitive" accounts of science in significant part because he tacitly identifies cognition with propositional attitudes toward an already determinate meaning or intentional content. Latour insists that what matters in science is not the thoughts of individual scientists but how their thoughts are "translated" into other terms (which include both straightforwardly linguistic translations and their realization in machines, pictures, and performances).[13]

Cartwright's work may clarify why the move toward an antiempiricist positivism also leads toward such broader criticisms of semantic realism.[14] Whereas early logical positivism had attempted to fix and legitimate the meaning of theoretical concepts by reducing the scope of theoretical claims to their empirical consequences, the postpositivist appeal to literal meaning countered reductionism with what Cartwright has aptly called a "fundamentalist" stance.[15] Fundamentalists take scientific theories and laws to apply everywhere, even to situations that no one knows how to formulate in terms of the theory. Such a fundamentalist stance overlooks the work that has to be done both in manipulating things and in redescribing them to make them amenable to theoretical interpretation (for as Cartwright notes, such redescription is prior to the application of the theory). The abstract formulation of a theory does not by itself specify how to apply it to a range of concrete situations. Instead, we must produce a model of the situation which would lend itself to interpretation in terms of the theory, and no guarantee exists that adequate models can be constructed for every possible situation that might prima facie be

12. Hacking 1983: chap. 1.

13. Latour sometimes speaks as if people's thoughts have quite determinate content independent of their public consequences, yet the content of thought is mostly irrelevant to the outcome of their actions; at other times he seems to take the intentional content of a thought to be determined only by what happens to it in subsequent interpretation and use.

14. The most familiar connection between these two concerns, of course, is the widespread recognition that the two dogmas of empiricism criticized by Quine (1953), analyticity and reductionism, are really the same dogma looked at in two different ways.

15. Cartwright 1994; note that the fundamentalist-reductionist divide actually predated the recognized demise of positivism, because Carnap, for example, eventually advocated a strongly fundamentalist position.

considered within the theory's domain. Cartwright argues against the fundamentalist that "the multitude of 'bad' cases, where the models, if available at all, provide a very poor image of the situation, are not cases that disconfirm the theory. You can't show that the predictions of a theory for a given situation are false until you have managed to describe the situation in the language of the theory. When the models are too bad a fit, the theory . . . is just inapplicable."[16] Yet Cartwright also rejects reductionism, because she quite reasonably wants to leave open the possibility of constructing new models that extend the same theory to other situations. The mistake common to both reductionists and fundamentalists is to presume that the scope and content of a theory are already fixed, either limited to the models already available or specified without limit by the meaning of the theory's abstract concepts.

Where Cartwright claims that theories apply only where they have reasonably accurate models, Hacking suggests more generally that "the very candidates for truth or falsehood have no existence independent of the styles of reasoning that settle what it is to be true or false in their domain."[17] Relating ostensibly different situations by constructing and manipulating families of analogically related models would be just one of many such styles of reasoning. Hacking's conception of a style of reasoning does not merely refer to patterns of thought, however, but also more generally to a whole range of practices within which those patterns could be intelligibly situated. Thus, practices of classification and counting are part of a statistical style of reasoning, whereas an experimental "style" presumably incorporates the development of instruments, laboratories, and procedures for controlling, monitoring, and registering the course of experiments. The point of connecting the truth-or-falsity of propositions to styles of reasoning is to suggest that whether those propositions say anything about the world, and what they turn out to say (whether true or false), depends on how they can be connected to things in the world via patterns of reasoning and the practices within which those patterns make sense.

Latour's development of a similar theme characteristically takes it beyond the realm of theories and propositions. He claims (as a "Rule of Method," no less) that "by themselves, a statement, a piece of machinery, a process are lost. By looking only at them and at their

16. Cartwright 1994: 283.
17. Hacking 1982: 49.

internal properties, you cannot decide if they are true or false, efficient or wasteful, costly or cheap, strong or frail. These characteristics are only gained through *incorporation* into other statements, processes and pieces of machinery."[18] Yet all three claims, Cartwright's and Hacking's as well as Latour's, can be understood to be within the spirit of the famous logical positivist doctrine that the meaning of a statement is its method of verification. Models, styles of reasoning, and incorporation into networks are all ways of establishing connections between a statement and publicly accessible things and events. The crucial difference (apart from the rejection of classically empiricist conceptions of manifestation) is that all three proposals abandon the idea that "meaning" in this sense is an already determinate property of a statement, instead seeing it as developed through dynamic and open interactions with the world. Thus, meanings themselves turn out to be dispensable reifications within the broadly positivist project of showing how scientific claims become meaningful only through explicit connections to public manifestations.

Recognizing the renewal of explicitly positivist themes in recent philosophy of science strongly suggests it is a misconception to take the recent history of the discipline as a struggle to identify and reject the fundamental errors of positivism. This insight will be especially compelling for those who admire, as I do, the spirit and achievements of philosophers such as Cartwright, Hacking, Fine, and Latour. Yet I regard the conception of philosophy of science as postpositivist to be not only an error but also a consequential error. This conception at times both caricatures the positivist movement itself and fails to grasp its constructive connections to more recent philosophical work. More significant are the ways in which the development of the field as a self-conscious reaction to positivism has blocked understanding and assimilation of some philosophically significant studies of science in other disciplines, most notably, I shall argue, cultural studies of science and feminist studies of gender in science.

The philosophical failure to recognize and to respond to cultural studies and gender studies is subtle but fundamental. To see the difficulty here, it is crucial to recall that an important and worthy theme in philosophy of science since the 1960s has been the acknowledgment of investigations of science by historians, psychologists, sociologists, and others and the blurring of any distinction between philosophical analysis and empirical studies of science. Philosophers

18. Latour 1987: 29.

readily made common cause with the intellectual historiography that came to them from Alexandre Koyré through Thomas Kuhn. Some philosophers have subsequently embraced and adapted the findings of cognitive psychology and neuroscience. Partly through the mediation of philosophically inclined scientists such as Richard Lewontin, others have brought their work into the conversation over conceptual issues within specific scientific fields. There has even been grudging acknowledgment (much of which has been highly critical) of the sociological studies that first emerged from Edinburgh and Bath in the mid-1970s and the subsequent European ethnographies of California laboratory life. This background accentuates the lack of a more widespread and serious philosophical response to a decade of feminist investigations of the natural sciences[19] and the similar silence toward the partially overlapping historical and anthropological work that has affinities with interdisciplinary cultural studies.

There might be a relatively old-fashioned sociological explanation for this gap, which would appeal to publication venues, citation patterns, and familiar delays in assimilating unfamiliar literatures. A different sort of explanation would emphasize the resistance of male-dominated fields to taking feminist scholarship seriously. I would like to suggest another alternative, however, with potentially more interesting implications. I suspect that an important reason why philosophers of science have had difficulty responding to cultural studies and gender studies is because these inquiries do not fit well into the stories with which we make sense of ourselves philosophically. In almost every important respect, cultural studies of science and investigations of gender and science pose challenges to the very terms in which philosophers of science have differentiated current work from logical positivism—or so I shall argue. To take these studies seriously threatens the very intelligibility of our disciplinary self-conception, which remains focused by the opposition to positivism.[20] This apparent threat might be interestingly turned around, however. Serious attention to cultural and gender studies might indicate how philosophical reflection on the sciences could constructively abandon

19. Feminist philosophers have, of course, both contributed and responded to the work on gender and science. What is striking is the limited response to their work by other philosophers of science, compared with the interpretive and critical response offered to other empirical research programs.

20. As I noted in Chapter 1, a frequently recurring motif in much philosophical work since 1970 (including some of my own) is the claim that earlier philosophical positions failed to recognize the most fundamental flaws in logical positivism and hence have not (yet) fully broken with the positivist legacy.

"postpositivism," so that differentiating our philosophical practice from logical positivism could at last cease to be a significant concern.

To understand this possibility, we need to consider in detail some of the ways in which cultural and gender studies might challenge philosophers' self-consciousness as postpositivist. I shall emphasize criticism of postpositivist conceptions of scientific communities, the epistemic primacy of global theory, the causal structure of reality, and the coalition of history and philosophy of science. Underlying all these issues, I suggest, is a challenge to what many philosophers (though not us alone) have seen to be at stake in postpositivist studies of science. Clearly, my primary goal in articulating such an intersection between philosophy of science and cultural and gender studies is an opening of philosophical horizons which might transform how we think about the sciences and about our own relation to the sciences. The essays in Part II offer one example of what philosophical recognition of and response to cultural and gender studies might contribute to understanding the sciences. But I also hope that further reflection on the philosophical challenges posed by cultural and gender studies of science might have interesting implications for cultural and gender studies more generally.

In considering the challenges to postpositivist conceptions of theory, community, reality, and history, I shall begin with scientific communities, in part because this concern accounts for some especially egregious misunderstandings of cultural and gender studies. Logical positivism understood science in terms of its formally specifiable methods, methods that could be grasped and applied by individual knowers, *any* individual knowers, because no philosophically relevant differences were thought to exist among them (except, perhaps, their willingness to engage in openminded, critical inquiry). The postpositivist turn suggested that there were no nontrivial and non-occlusive methods of inquiry without far-reaching substantive presuppositions about the world investigated.[21] Such presuppositions formed the basis for communities of inquiry: the intelligibility of their shared project depended on their acceptance of the fundamental presuppositions underlying their methods. The scientific or epistemic communities recognized by postpositivist philosophy of science are thus consensus communities, defined by shared beliefs, values, standards, and projects.

21. As this paragraph indicates, it is difficult to talk about any of the fundamental themes of postpositivism without also invoking the others. This interconnectedness has increased the difficulty of a postpositivist philosophical encounter with cultural and gender studies of science.

Cultural and gender studies generally reject the postpositivist picture of relatively autonomous scientific communities identified by shared networks of beliefs, values, and methods, but not in order to return to the sovereignty of the individual knower. Groups, networks, social or political alignments, and subcultures remain of fundamental importance to cultural studies, but the notion that consensus defines identity has been eroded. There is no place within most postpositivist philosophical perspectives for the emphasis of cultural and gender studies on fragmented or contested identities or on heterogeneous alignments or solidarities that do not reduce to either shared beliefs and values or tolerance for individual differences. This point may help account for the puzzlement with which some "mainstream" philosophers of science respond to feminist studies of gender in science: analyses of the gendering of science would be puzzling if they identified women as an undifferentiated epistemic community united by common interests, values, and beliefs that are supposed to be explicitly rejected or devalued by the beliefs and practices of (male-dominated) scientific communities. Only the most blatant forms of gender bias and domination are even recognizable in these terms, and this fact has enabled many philosophers of science to misread or dismiss more subtle and far-reaching analyses.

The criticism of the presumed autonomy of scientific communities may be even more consequential for a postpositivist philosophical self-conception. With the demise of the positivist account of formal methods that underlie unified science, the epistemic and political autonomy of scientific communities becomes the basis for identifying "science" as a distinctive and relatively confined topic of philosophical reflection. The problem is not that philosophers fail to recognize that the autonomy of scientific communities is open to challenge; rather, we often do not discern the depth of the challenge. Understanding the question in terms of "external" influences on scientific communities presupposes clear boundaries between scientific communities and the social, cultural, or political "contexts" that influence them. This interpretation of the criticism mistakenly reifies scientific communities and the predetermined social "factors" that might influence them. But this interpretation also makes invisible a distinctive contribution from cultural and gender studies, which at their best examine the dynamics of communities, identities, and boundaries. They address the ways in which meaningful differences and boundaries are established, sustained, and transformed; the ways they fragment; and the tensions and resonances which indicate that apparently stable identities and differences are sustained only by ongoing work.

Feminist studies of gender have often been in the forefront on this issue, as gender has been increasingly recognized not as something given but as a field of possibilities, sustained, differentiated, and challenged by ongoing performances.[22] Studies of gender and science have deepened this realization by problematizing the original distinction of gender from sex, which had identified gender as a cultural construction founded on something given biologically. Nevertheless, a wide range of other cultural studies of science also resist categorization in terms of what is inside and what is outside: in addition to gender studies, I think of Donna Haraway on the reconfiguring of biology in terms of cybernetic systems;[23] of Robert Marc Friedman on the construction of a modern meteorology;[24] of Sharon Traweek on multiple forms of marginality as epistemic resources and liabilities in physics;[25] of Mario Biagioli on the Galilean legitimation of mathematics as natural philosophy within the Medici court;[26] of Vassiliki Betty Smocovitis on the evolutionary synthesis as "unifying biology";[27] or of Paula Treichler and others on AIDS activism and the shaping of medical discourse and research protocols.[28] Others familiar with cultural studies of science could undoubtedly expand this list quite readily.

Cultural and gender studies, then, take us away from the postpositivist focus on scientific communities and the shared content of presuppositions constitutive of their research. Epistemically significant groups are defined not by the common possession of some presupposed content but by their belonging to partially shared situations, to a cultural field. "Science" and scientific knowledge are not framed by the identification of scientific communities but are among the meanings at issue in the formation and interaction of various cultural alignments and groups.

Cultural and gender studies, however, also pose deeper criticisms of the postpositivist turn to the content of global theories as a replacement for positivist conceptions of scientific method. As I noted earlier, the logical positivists were suspicious of the meaningfulness of theoretical discourse; they initially suggested that the meaning of theoretical statements is confined to their empirically or operationally

22. See esp. Butler 1990, 1993.
23. Haraway 1981–82.
24. Friedman 1989.
25. Traweek 1992.
26. Biagioli 1993.
27. Smocovitis 1992.
28. Treichler 1988, 1992; Patton 1990.

specifiable consequences. Postpositivist philosophy of science has inclined toward rather more robust conceptions of theoretical understanding: far from being subordinate to scientists' empirical capabilities, scientific theories may encompass global "worldviews," providing the context within which specific interpretations, measurements, or observations could make sense. Furthermore, theories provide such far-reaching interpretations simply by virtue of what they say about the world. Van Fraassen is among the more prominent dissenters from the general trend toward scientific realism, yet even he enunciates a more widespread tendency in contemporary philosophy of science when he differentiates his own empiricism from logical positivism by "insisting upon a literal construal of the language of science, [which] rules out the construal of a theory as a metaphor or simile, or as intelligible only after it is 'demythologized' or subjected to some other sort of 'translation.' "[29]

When practitioners of cultural studies object to the notion of a straightforwardly literal construal of scientific theories, they do so not to substitute a preferred nonliteral one. Neither do they share the (post-)positivist preoccupation with the status of "theory" as a unified and distinctive kind of semantic structure. They offer instead a richer and more fluid understanding of the signifying possibilities of the discourses within which scientific practices are situated. Their focus shifts to occasioned utterances and other performances, which are significant not as tokens of conventionally fixed meaning-types but only as mappings of various signifying acts onto others. Put another way, there is no propositional content intervening between occasioned utterances and the myriad ways they can be taken up and used on subsequent occasions (along with the ways they themselves appropriated previous significations). The interesting question is not which utterances truly describe the world, *given* a "literal" understanding of what those utterances say, but how utterances come to be significant sayings (that is, intelligible, worth mentioning, and perhaps even "serious" or important) in the light of what is said on other occasions before and after. This shift comes about not because cultural studies divorce language from the world but because what people say acquires its significance in part through the ways they encounter and respond to things. There is no prelinguistic access to the world, but then neither is there any otherworldly access to linguistic content, to what sentences literally say. These concerns

29. Van Fraassen 1980: 11.

about the reification of semantic content thus have important and obvious affinities to the work of Cartwright, Hacking, and Latour discussed above. Cultural studies would join Cartwright in rejecting the dichotomy between fundamentalism and reductionism and insist on the openness of scientific significations. Theoretical representations cannot magically extend their application beyond the experimental and instrumental practices through which they become mutually adapted to a domain of phenomena, but neither can they be reduced to a fixed domain of significations. The range of a science's signifying possibilities is always open-ended (and possibly contested) and is partly what is at issue in the shaping of ongoing scientific research.

Cultural and gender studies also raise other concerns about the postpositivist philosophical turn toward global theoretical frameworks that condition all scientific understanding and practice. The problem is partly that scientific theory does not wear its meaning on its sleeve, so to speak,[30] and thus is meaningful only through its interrelations with other occasioned utterances. But cultural and gender studies would also contest the privileged position that much of postpositivist philosophy of science has assigned to networks of theoretical sentences. Hacking has already noted that postpositivist invocations of theoretical networks both overreach and homogenize. 'Theory' instead denotes a motley of things, including speculations, physical analogies, ad hoc mathematics, and the variety of models that articulate the world theoretically and make it calculable; much of scientific work also proceeds relatively independently of this motley of theorizings.[31] Cultural and gender studies strongly reinforce and extend Hacking's point. Although the manifold ways of theorizing are important and indispensable aspects of the meaning-making that happens through scientific work, it is also crucial not to neglect the ways they interconnect with narrative traditions, imagery, cultural practices and institutions, instruments (and material culture more generally), social relations, or affective responses and determinations if we are to understand how the sciences participate in signification.

These challenges to recent philosophical conceptions of the scope and the autonomy of theoretical understanding also undercut our sense of having rejected the positivist separation of facts and values. Postpositivist philosophy has frequently emphasized that there are

30. Indeed, strictly speaking, it has no meaning to wear, except through its interrelations with other signifyings; as an isolated sign, it is arbitrary and meaningless.

31. Hacking 1983: chap. 12, and 1992.

always evaluative presuppositions shaping scientific research and its outcome which are an integral aspect of the theoretical background that guides research practice for a particular scientific community or within a domain of inquiry. Yet this conception still reifies values as something possessed by a knower (in much the same ways that someone can also be said to possess desires or interests)[32] and also understates our own participation in or belonging to the evaluative concerns through which scientific work becomes significant. Cultural and gender studies suggest that such a conception omits from consideration the ways in which scientific practices and beliefs are situated within larger patterns of social interaction which distribute and sustain differences in power. These patterns include those that are sustained and reproduced in part by the production and certification of knowledge itself. Scientific research and development substantially affect the configuration of practices and meanings within which people's actions make sense to themselves and to one another.[33] Within those configurations, people differ significantly in the ways they are represented and acted on and the ways in which it is possible, or recognizable, for them to act in response (including speech acts); such differences are in turn enormously important for one's possibilities and prospects for a flourishing life.

So far, I have discussed two ways in which cultural and gender studies criticize self-consciously antipositivist themes in recent philosophy of science: the emphasis on scientific communities and on the theoretical presuppositions of their research programs. Both criticisms join the philosophical renewals of positivism in objecting to the notion of a common and fixed meaning or content expressed in scientific discourse and practices as literally interpreted. But identifying this concern with positivist themes may seem seriously misplaced. After all, some of the most thoroughgoing scientific realists among contemporary philosophers of science offer what initially seems to be a similarly radical criticism of the idea of shared content and its underlying conception of scientific rationality. Happy to accept "the fragmentation of reason" and a Feyerabendian anarchy about method,[34] these realists insist on the radical contingency of scientific success: scientific communities do not manage to describe

32. Indeed, I perceive widespread philosophical sympathy for something like Laudan's 1984 suggestion that theoretical, "axiological," and methodological presuppositions can be clearly distinguished and shown to vary independently.

33. An extended argument for why these effects should be understood in terms of power relations is developed in Rouse 1987b, especially chapters 6–7.

34. Stich 1990 and Churchland 1992.

the real world accurately because they developed rational methods; their methods are rational only to the extent that they have happened to hook onto a vocabulary that limned the real.[35] Yet, for scientific realists, meaning-content is still fixed by the chains of causal interactions in which that vocabulary is introduced; such an account is intelligible only if the sense and reference of 'cause' is fixed by more traditional means.

Cultural studies would reject such a scientific realism, but not because they betray an idealist or textualist disdain for reality or a skeptical suspicion that the real must always escape our comprehension. Indeed, cultural studies typically insist on the materiality of meaning-making and the indispensability of scientific interaction with and intervention on the world. Cultural studies and gender studies alike invite suspicion of the productionist and constructivist metaphors that suggest that "nature" or "the world" can denote nothing more than what some collectively powerful, gendered, and gendering "we" make it or take it to be. Cultural studies instead challenge the residual semantic realism with which even scientific realists connect things to words. Underlying this challenge is a quasi-nominalist[36] concern with how things become significant, with how their interconnections become manifest and their features become prominent in ways that matter.

The postpositivist turn to realism and to the semantic realism that promises literal interpretation of scientific language has been represented as a repudiation of the Vienna Circle's antirealism. Yet we have seen how the more familiar philosophical and sociological antirealisms still retain a realist core: interpretation must come to an end somewhere, if not in a world of independently real objects and their causal properties, then in the real structure of experience, a language or conceptual scheme, or a social context (perhaps a culture). Cultural studies instead promise a more thoroughgoing antirealism, which gives up altogether the dualism of scheme and content or context and content.[37] They still identify the real with what can become manifest,

35. See, for example, Boyd 1988.

36. I use "quasi-nominalist" because sayings and namings are not imposed full-blooded on things from "outside" but are partially shaped by the ways people interact with the things named and talked about.

37. As both Davidson and Latour have noted, such a thoroughgoing "antirealism" undercuts the invidious contrast between the things around us and the supposed real basis for their manifestation (experience, language, or culture) and thus does not have the air of paradox often associated with antirealism. Davidson (1984) states, "In giving up the dualism of scheme and world, we do not give up the world, but re-establish unmediated touch with the familiar objects whose antics make our sentences and opinions true or false" (198); and Latour (1988), "A little relativism takes one away from realism; a lot brings one back" (173).

that is, with what can possibly count as real within ongoing cultural practice, but what can count as ongoing cultural practice is treated similarly. Such an understanding does not even make prior commitments concerning which "subjects" can participate in such practices or on what terms they could participate. It can readily accommodate, for example, Haraway's reflections on Barbara Noske's *Humans and Other Animals,* which pose the task of engaging animals in nonreductive, "otherworldly" conversational practices that do not designate spokespersons for them. Haraway suggests that "we need other terms of conversation with animals. . . . The point is not new representations, but new *practices,* other forms of life rejoining humans and not-humans."[38] The practices she envisages could not be "our" constructions, even though people must actively participate in them. Such practices would reconfigure the "space" within which people and other beings show up and the terms in or on which we might be intelligible to one another, perhaps even including what intelligibility itself could consist in and for whom.

Such an account of how beings become manifest through significant practices historicizes the real without relativizing it (for there is nothing for it be relative *to*). The appeal to history might therefore seem to be one point of continuity between cultural studies and the self-consciously postpositivist stance of contemporary philosophy of science. Yet such convergence is belied by the ways in which histories of science have been significant within the postpositivist philosophical tradition. For philosophers of science, history has had two quite different roles, as archive and as grand legitimating narrative. In the former guise, history provides a repository of facts to which philosophical interpretations of science must be responsible. Kuhn canonically initiated this role for history when he suggested that "history could decisively transform the [positivist] image of science by which we are now possessed, if viewed as a repository for *more than* anecdote or chronology."[39] This initially liberating exposure of the historical disconnection of positivist accounts of scientific knowledge has increasingly been reified. Thus, some prominent philosophers have argued that the history of science now offers a collection of data which is sufficiently uncontaminated that the "empirical import" of various philosophical accounts of scientific change can be effectively tested against that historical record.[40] This odd conjunction of self-described postpositivist philosophy of science with a more nearly positivist phi-

38. Haraway 1992a: 87.
39. Kuhn 1970: 1 (my emphasis and rearrangement).
40. Laudan et al. 1986.

losophy of history is clearly and strikingly at odds with the historiographic orientation of cultural studies and, indeed, of cultural history more generally.

Yet the most significant barrier to philosophical recognition of and response to cultural and gender studies may be the predilection within much philosophy of science for the grand historical narratives of the legitimation project, despite the infrequency with which terms such as 'narrative', 'metanarrative', or 'legitimation' actually appear within the philosophical literature. The importance of modernist narratives within recent philosophy of science has already been discussed extensively in earlier chapters. But the resulting barrier to philosophical recognition of cultural and gender studies is not that many philosophers of science still want sweeping historical narratives of the progressive emergence of scientific rationality or truth, whereas cultural and gender studies would give them up. David Bloor, Harry Collins, and Barry Barnes certainly object to grand historical narratives legitimating the rationality or approximate truth of current science, and philosophers have given their work serious critical attention. The problem instead is that philosophers have recognized and responded only to works for which the adequacy of such legitimating narratives is an important question. Bloor's, Barnes's, and Collins's criticisms harbor a deeper agreement with many philosophers of science: that the cultural role of modern sciences is in need of such global legitimation. These sociologists all take it to be a consequential discovery that the outcome of scientific work cannot be explained in terms of its rationality or truth. Thus, we see once again that the horizons of philosophy of science have been fundamentally shaped by the legitimation project.

Cultural studies and gender studies are therefore off the conceptual map for many philosophers of science not because they reject realism or the emergent rationality of science but because they do not engage the questions to which these views are posed as answers. I believe that by transcending these questions, cultural and gender studies offer a significant opportunity for a redirection and renewal of philosophical reflections on the sciences. I would thus like to conclude by suggesting some consequences of abandoning the project of defending or criticizing grand historical legitimations of scientific knowledge and thereby of taking seriously the contributions of cultural and gender studies to a renewal of philosophical reflection on sciences and knowledge. Many of these themes will be developed in the following chapters.

Cultural studies and gender studies typically focus their attention on what is at issue in practices that are situated: historically, geographically, and culturally. Such practices belong to a meaningfully configured world that allows specific possible ways to be, act, and signify to emerge as significant for the beings that it encompasses. Its configurations are not outrightly hegemonic, for they are always shaped in part by the ways their emerging patterns are resisted and reappropriated: they establish not static structures of meaning but dynamic situations. Examining such configurations of meaningful possibilities may nevertheless be more fundamental to understanding knowledge making than are the questions about truth, justification, and explanation which have been so central to philosophers of science: these examinations would address what is at issue in settling a truth claim, how that issue comes to be seen as intelligible and important, and what is at stake in its settlement and for whom.

Cultural and gender studies would also introduce at least two important topical shifts into postpositivist philosophical reflections on the sciences. The first shift concerns the disciplinary location of philosophical attention to science. Despite frequently renouncing the project of demarcating scientific inquiry from other kinds of activities or belief structures, philosophers of science have continued to focus their attention on elite, academic sciences, ones whose historical construction as disciplines has been closely tied to ideals of a pure science unsullied by practical, material, or political pursuits.[41] One could read the professional journals and the most prominent books in the philosophy of science for a considerable time and rarely encounter the notion that science has anything to do with medicine, agriculture, sex, industry, or war. Instead, philosophers of science attend almost exclusively to physics, astronomy, biology, physical and inorganic chemistry, and cognitive psychology. Even within these fields, philosophical attention has been directed toward those prestigious, theoretically ambitious subfields that most promise to secure the unity and autonomy of scientific work (for example, quantum mechanics, relativity theory, and high-energy physics or evolutionary biology and molecular genetics). When philosophers do explore the boundaries

41. The conception of such scientific fields as quantum physics, evolutionary biology, or molecular genetics as pure, academic sciences disengaged from applications or political controversy is rather disingenuous, given the Manhattan Project and the complex political struggles over evolution, eugenics, and biotechnology. Nevertheless, philosophical discussion of these disciplines has effectively treated them as highly autonomous fields, focusing on their internal theoretical structure and their capacities for explanation and unification.

around these high-profile examples of rarified academic science, they typically consider fields such as astrology or parapsychology, whose marginality to elite science can be understood philosophically in terms of pretense to the status of academic science and which can therefore help to secure those boundaries. Cultural studies and gender studies have been much less inclined to police the borders of the sciences and rather more attentive to the constant traffic across them. In thus noticing how thoroughly scientific practices and scientific authority are intertwined with contemporary politics and culture, they also suggest a profound shift in the issues and practices that might be of interest to a philosophy of science.

The other topical shift that cultural and gender studies would introduce undercuts the exceptional significance of history for the construction of postpositivist philosophies of science. For philosophers, histories of science introduced difference into the positivist dream of a methodologically unified science. But apart from recognizing some irreducible differences among scientific disciplines, philosophers have largely consigned significant differences within the sciences to their past.[42] Such consignment has been strategically important for constructing narratives of the progressive emergence of scientific rationality or truth. Historical time is not, however, a similarly privileged site of difference within cultural and gender studies. They therefore invite philosophical attention to other overlapping dimensions along which contemporary scientific practice constructs or displays significant identities and differences: geography, power, sex/gender, race, nationality/culture, species, institutions.

These two topical shifts promise in turn a renewal of the political engagement of philosophical reflection on the sciences. For the positivists (and for their contemporary Karl Popper), philosophy of science was the focus of a much broader program of political and cultural reform. Postpositivist philosophy of science has reacted strongly against such political engagement. Many sociologists still criticize the supposedly "normative" stance of philosophy of science, but as Steve Fuller has argued persuasively, its evaluative concerns are largely retrospective and aesthetic.[43] Recent philosophy of science may propose the legitimation of current scientific practice, but it provides minimal resources for its criticism or transformation. Indeed, in an assessment of the significance of Thomas Kuhn's work, Fuller

42. A striking exception to this trend is Dupre 1993.
43. See esp. Fuller 1989: 9–10.

has noted how deeply postpositivist philosophy of science is rooted in the conception of "the end of ideology" which flourished in the United States in the 1950s and 1960s.[44] Cultural studies and gender studies suggest rather different conceptions of political criticism than do the positivists' modernist manifestoes, but they share with positivism a commitment to a politically engaged practice, a commitment that we philosophers of science might do well to recover for ourselves.

Apart from philosophers' parochial concern to sustain and renew an inherited discipline, why should it matter to reconstruct the stories that have shaped a postpositivist philosophy of science, so as to allow for such an intersection with cultural studies and gender studies? What significance could accrue to such a retelling that would seek to situate the sciences historically and culturally, to open their borders rather than secure their closure, and to engage the sciences philosophically without pretense to political innocence? These questions go to the heart of this book, because for all my concern to open the carefully policed borders of philosophy of science, many readers will recognize the style and focus of my concerns as unmistakably philosophical. I would also be the last to deny the grip that the postpositivist stories I want to reconstruct still have on my own philosophical imagination. Disciplinary boundaries remain important, however more fluid and nonexclusive they might become. Disciplines, after all, have histories and cultural locations that can be challenged and reconfigured but not simply surpassed. A successful reconstruction must therefore do more than merely supplant philosophy of science with cultural studies, or colonize cultural studies and gender studies for previously defined philosophical purposes. It must open a space for genuine and ongoing conversation—a much more fragile but in the end possibly far more satisfying outcome. The upcoming chapters are my opening contribution to such a possible conversation.

44. Fuller 1992a.

Part II

RECONSTRUCTING PHILOSOPHY OF SCIENCE: PRACTICES AND THE DYNAMICS OF KNOWING

[5]

The Significance of Scientific Practices

In the first part of this book, I criticized the commitment of philosophy and sociology of science to the global legitimation or criticism of scientific knowledge and the cultural authority of science. I also objected to the anxious preoccupation with positivism still reflected in the disciplinary self-conception of philosophy of science as "postpositivist." Such criticism rings hollow, however, unless we can see how science studies might get out of the legitimation business and overcome our anxiety about vestigial positivism. This part of the book points toward cultural studies as a promising alternative model for science studies. A first step toward understanding science studies as cultural studies, however, involves further reflection on the conception of science as *practice*. This conception was a central theme of my earlier book, *Knowledge and Power*,[1] and the notion of practice, or social practice, has been increasingly prominent in other studies of science as well. Unfortunately, the increasingly frequent invocations of science as "practice" conceal equivocations that point toward divergent philosophical agendas. Some clarification and criticism is thus necessary to understand how cultural studies of science might emerge from my account of scientific practice.

My account must be carefully differentiated from at least three other ways of understanding science as practice. The differentiation

1. Rouse 1987b.

will be subtle, for important affinities to my account also exist. Nevertheless, crucial differences emerge over the philosophical significance of understanding science as practice. I begin with the most philosophically articulated conceptions of practice, which share an underlying commitment to a kind of humanism. In different ways, Hubert Dreyfus, Robert Brandom, and Peter Winch have attempted to reinterpret human subjectivity in terms of practices rather than beliefs and desires.[2] The aim of these reinterpretations has been to provide more adequate justifications for philosophical distinctions between freedom and nature or between the human sciences and the natural sciences. My account of scientific practices must also be carefully distinguished from social constructivist accounts of agency and practice. Social constructivists emphasize the distinctive importance of social practices in situating and authenticating scientific knowledge in order to justify the primacy of social explanations of the outcome of scientific work. The most prominent social constructivists have important philosophical affinities with some of the philosophical "humanists" I discuss, even though the latter all resist fully incorporating nature and natural science within their accounts of practices. Finally, it is important not to confuse my account with the Marxian or Braudelian materialist themes that emerge in some work in science studies, emphasizing the primacy of a material "basis" of science over its more superficial discursive articulation. Materialist conceptions of scientific practice have their origins in the welcome renewed historical and philosophical attention to experiment and instrumentation in the sciences. But such conceptions take this emphasis much further, through a philosophical or historiographical reduction of discursive practices to "literary technology" or the production of rhetorical force.

To understand how my account of scientific practices diverges from these humanist, social constructivist, and materialist approaches, it may be useful to recapitulate some of the central themes of my initial treatment in *Knowledge and Power* of science as practice. I will also foreground some of the concrete features of scientific work that emerge as central to any account that emphasizes scientific practices. In subsequent sections of the chapter, I will return to the concept

2. Dreyfus 1980, 1984, 1991; Brandom 1979, 1994, 1976; Winch 1958, 1964. Pickering (1995) has more recently developed a conception of practice that has important affinities to mine and shares my reservations about commitments to humanism, but the full articulation of this account appeared too late for me to undertake here a thorough assessment of the importance of its similarities and differences from what follows. Brandom 1994 also appeared too late for me to assess the extent which his (1979) views have changed in ways that would affect my argument.

of practice, first to clarify how my account differs from these other approaches and then to discuss the significance of conceiving scientific practice in this way.

SCIENTIFIC PRACTICES

In *Knowledge and Power* I focused on the notion of science as practice primarily as an alternative to the epistemological or representational conceptions of science which strongly motivate the legitimation project. Often, practice is opposed to theory. That opposition may seem reinforced by the common philosophical emphasis on scientific theories as the most important and far-reaching instances of scientific knowledge. But my conception of science as practice did not distinguish practice from theory; theorizing is as much a practice as any other aspect of scientific work. Instead, I emphasized scientific research as a kind of practical activity, one that reconstructed the world as well as redescribed it.

Theoretical models and experimental microworlds were the two foci of scientific work which stood out most prominently in my earlier description of scientific practice. Scientists construct and manipulate a variety of abstract and idealized models.[3] Such models take quite disparate forms: mathematical, schematic, verbal, and even physical. They do not combine to form a systematically unified representation of the world. Instead, they exemplify "a picture of theory as a disjoint collection of models whose range of application is not fully specified and whose effectiveness and accuracy vary considerably within that range. . . . Their coverage overlaps, and they may provide inconsistent versions of the same phenomenon. Some phenomena may fall in the gaps between the various kinds of [models] we have in a domain and are consequently not well treated by any."[4] Models are significant for what one can do with them: the operations that can be performed

3. Cartwright (1989: 185–98) has emphasized an important distinction between idealization and abstraction in the formulation of theoretical models in science. Idealizations are concrete (but often fictional) situations whose causally relevant parameters are carefully chosen for ease of calculation or clarity of presentation (frictionless planes, perfectly elastic collisions, and two-particle universes are thus idealizations). Abstractions are *general* models that omit some causally relevant factors in order to display patterns or tendencies that are manifest across a range of concrete situations (Galileo's law of free fall, Boyle's Law, and the quantum mechanical Hamiltonian for a hydrogen atom are abstractions). The two are closely connected: the idealized model often indicates the patterns that most clearly display an abstract law, but they differ in their specificity and their range of applicability.

4. Rouse 1987b: 85.

on or with them, their response to various operations and perturbations, and the range of phenomena to which they can be adapted with appropriate phenomenological adjustments.[5] This conception of theoretical modeling is not strictly instrumentalist, even though most scientific models are adapted to situations they do not literally describe; it is consistent with accepting the existence of theoretical entities and allows for the possibility that some theoretical models realistically describe some concrete situations (and not merely the "observable" ones).[6]

Microworlds are local reconstructions of the world to enable more effective manipulation and control and more careful and revealing surveillance of outcomes. Laboratories are the characteristic site for the construction of microworlds, although scientists perform similar reconstructions in field studies and other settings.[7] Experimental microworlds are settings that allow interactions among selected objects to be brought about and closely monitored under conditions of causal isolation or randomization (other objects or events that might affect the outcome of the intended interactions are either excluded from the setting of the microworld, or their influence is deliberately randomized to prevent any systematic effect). The significance of microworld interactions often results from decisive changes in the scale of interaction among their constituents and with the experimenters.[8] These microworld constituents often include objects that are created or isolated only in that setting. More generally, their constituents are purified, measured, segregated, and enclosed to enable them to manifest themselves in ways that would not show up clearly else-

5. The principal sources for my conception of theories as families of models are Nancy Cartwright, Ronald Giere, Thomas Kuhn, and Ian Hacking. Nevertheless, my resulting conception of theorizing as a practice that engages with and transforms the world is closer to Bruno Latour's contemporaneous discussion of "centers of calculation" (1987: chap. 6).

6. Cartwright (1994) has clarified in a useful way the relationship between this conception of theorizing as modeling and the debates over realism and instrumentalism. The target in conceiving of theories as collections of models is not realism but "fundamentalism," the view that scientific laws and theories have an unlimited scope. The fundamentalist takes models to articulate the content of a theory; for the antifundamentalist, the accurately constructable models open-endedly constitute what the theory has to say about the world. Rouse (1987b: 148–52) anticipates Cartwright's reformulation, arguing that her scruples about the range of precise application of fundamental theories are consistent with such theories being true, if 'truth' is understood in a suitably deflationary sense.

7. Field research also often establishes a complex traffic between field sites and laboratories, as objects and signs extracted from the field are introduced into microworld settings in laboratories, the outcome of which may then be reconnected to the field site in various ways.

8. Latour (1987: chap. 6, and 1983) is especially eloquent on the importance of changing scale for establishing a more effective scientific practice.

where. These manifestations themselves typically depend on locally integrated networks of instrumentation, technical skill, and object signification. This last term indicates an especially important but frequently overlooked aspect of experimental practice: inducing objects to produce signs. In *Knowledge and Power,* I noted the examples of radioactive labeling, cloud and bubble chambers, X-ray crystallography, and the various forms of chromatography, spectroscopy, microscopy, and telescopy, but the inventiveness of scientific work has proliferated such forced signification well beyond these classic cases.

I speak of experimental microworlds rather than simply of experiments because experimental practice is typically organized around the family of objects, settings, and controls that constitute the microworld. As François Jacob once noted: "In analyzing a problem, the biologist is constrained to focus on a fragment of reality, on a piece of the universe which he arbitrarily isolates to define certain of its parameters. In biology, any study thus begins with the choice of a 'system'. On this choice depend the experimenter's freedom to maneuver, the nature of the questions he is free to ask, and even, often, the type of answer he can obtain."[9] The establishment of such an experimental system or microworld permits a sustained program of research whose focus is not just the specific results of individual experiments but a better understanding of the entire experimental system as a setting for ongoing research. Often, especially informative microworlds are established as objects of study by multiple groups of researchers over extended periods of time (whether by sharing the same facility or by reproducing the system elsewhere). Thus, for example, the chromosomal and phenotypical variations of crossbred *Drosophila melanogaster,* the eighty-two-inch bubble chamber at Stanford Linear Accelerator, and the behavioral interactions of chimpanzees at the Gombe field station became important "locations" for sustained study, not just because of their intrinsic characteristics, but also in large part because of the historically accumulating "background knowledge" and the associated institutional facilitation of access and protocol.

Experimental work places a premium on introducing and monitoring controlled disturbances into previously stable and well-understood settings. The clarity and significance of such practices depend on the establishment of both a stable ground state for the microworld

9. Jacob 1981: 234; I am grateful to Hans-Jörg Rheinberger for calling my attention to this passage.

setting and a more thorough and rich understanding of its constituents and their capacities, tendencies, and possibilities. Thus, experimenters need to know in a general way the constituents of an experimental system and their expected kinds of interaction as well as to have a detailed practical familiarity with the system's constituents. Such practical understanding enables a sensitivity to when the system is working "properly" and to when unexpected variations are significant and when they are just noise to be eliminated. It also enables the ongoing adaptation of the microworld to local opportunities, protocols, and needs. At the same time, it encourages the packaging and standardization of instruments, techniques, protocols, and even whole microworlds both to enable their transfer and adaptation to new settings and to build in the avoidance of previously recognized pitfalls and problems without having to reinvent their local circumvention.

Scientific work involves multiple interactions and overlaps between theoretical modeling and the deployment of experimental microworlds. The idealizations and abstractions embedded within models may guide the construction and the characterization of microworlds, whereas microworlds may suggest directions for the development, elaboration, and interpretation of models. To some extent, the availability of a good experimental system may itself be a motivation for modeling it theoretically, and informative theoretical models may provide targets for experimental construction. Nevertheless, we should not overinterpret the relationships between them. Theoretical models often embody conditions that are unrealizable (sometimes in principle and sometimes as a matter of practical constraints on materials or funds), whereas the actual construction of experimental systems often involves levels of specificity and local contingency which do not readily fit into models. Perhaps more important, theory and experiment are developed in response to many considerations other than just their mutual articulation. There are distinctively experimental, instrumental, or theoretical opportunities, questions, problems, and constraints to which these practices respond. Moreover, scientific practices are differentially influenced by their various economic, cultural, and political engagements. Thus we should see models and microworlds not as directed toward the telos of a unified practice of material and symbolic abstraction but as more dispersed and centrifugal engagements with the world.

An important theme of *Knowledge and Power* concerned how abstract and idealized models and the cloistered microworlds of experimental practice are connected to the more muddled and entangled things and events outside the research setting. In part, one accomplishes this connection by adapting the theoretical models and standardizing and desensitizing laboratory practices so that they are useful in less carefully constrained circumstances. The models are applied with less precision and specificity, or their ceteris paribus clauses are filled in phenomenologically. Instruments, procedures, and techniques are modified to permit less skillful use and to be less responsive to changing conditions of application. Precision and discrimination are typically sacrificed for the sake of robustness.

More crucial to the extension of scientific practices beyond the research setting, however, is the reconstruction of the surrounding world to resemble the laboratory in important respects. Objects and substances created in or for the laboratory are introduced into other settings. Partitions and enclosures are built to prevent unwanted or unaccountable mixtures. Actions and events are more carefully sequenced and timed. Instruments to register and interpret the signs first elicited from objects in laboratories become standard equipment elsewhere. Measurements become a more constant feature of everyday practice, while the units of measure and their application are increasingly standardized across widely variant contexts. The extension and commensuration of measures is by itself an enormous ongoing effort to sustain the wider extension of scientific practices and understanding. Meanwhile, all these transformations are supported by more extensive and careful surveillance and tracking of their outcomes and by elaborate documentation and accounting.

This ongoing transformation of the world to make it more readily and extensively knowable also imposes changes on people. Partly this imposition is a matter of discipline. If objects and events are to be more carefully regimented, sequestered, and timed, people's actions must also be more carefully monitored and controlled. There is less room for either inattention or independence, and there are more specific constraints on what people do. As Langdon Winner once noted, "Certain kinds of regularized service must be rendered to an instrument before it has any utility at all. One must be aware of the patterns of behavior demanded of the individual or of society in order to accommodate the instrument within the life process."[10] Acquisition, supply, operation, and maintenance are necessary not just for specific instruments but also more generally for the larger equip-

10. Winner 1977: 194–95.

mental contexts through which scientifically constructed microworlds are connected to what happens elsewhere.

The disciplines needed to establish and extend laboratory practices and achievements also include the habitual practices and skills through which people make themselves into competent, reliable participants in a more or less shared world. Who we are is in significant part whom we have made ourselves into through the cultivation of habits of mind and body. A familiar and pervasive example is literacy. The ability to read and to write transforms us not only through the self-control needed to acquire it and the stock of knowledge it makes available but also through the associated transformation of patterns of thinking and of what can be recognized as excellence of thought.[11] Scientific practices include specific extensions of literacy, of course, but they also involve many other transformations of self. Mathematical competence (in quite specific forms: the calculus, statistical reasoning, logic, abstract algebra, and nonlinear mathematics form us as thinkers in somewhat different ways), perceptual focus and discrimination (natural history and *Drosophila* genetics, for example, develop people's perceptual recognitions in quite distinct ways), and a skillful grasp of particular instruments are specific and sometimes far-reaching transformations of self. Furthermore, these practices are typically situated in specific social and cultural contexts; acquiring a physicist's understanding of the world is not fully or readily separable from other aspects of one's socialization as a physicist.[12]

Perhaps the most important effects of scientific practices on how people live are more indirect, however. The situations in which people find themselves and the actions, roles, and interpretations through which they respond to situations can make sense only within a historically and culturally specific configuration of meaningful possibilities. Such contexts enable us to understand concretely who we might be, what we might live for, how we can meaningfully interact with others, what equipment and procedures we can utilize, and how we encounter our surroundings. The practices that have come to be called "scientific" have been integral to the remaking of these fields of possibility globally, not merely in the "West."[13] Scientific practices, and the extension of their models, practices, and constituents beyond

11. Goody 1977.

12. Traweek 1988.

13. Indeed, scientific and technological practices have importantly shaped the cultural self-conception that defines itself as "Western"; for a useful historical discussion of this issue, see Adas 1989.

the laboratory, reconfigure the possibilities in terms of which people can intelligibly understand and enact their lives.

PRACTICES: THE VERY IDEA

Understanding scientific work as practices in this way requires significant revisions both in familiar ways of thinking about scientific knowledge and in how the notion of a "practice" has figured in contemporary philosophical discussion. In this chapter I focus on practices, deferring most of my discussion of scientific knowledge until later. Nevertheless, my interpretation of practices is motivated in significant part by its implications for understanding scientific knowledge. In particular, my objections to "humanist" accounts of practices importantly turn on my concerns about scientific knowledge. Hence, my initial reflections on the notion of a practice must include some consideration of the account of knowledge that will emerge in what follows.

We are accustomed to thinking of knowledge as something possessed by a knower. Knowers might be either individuals or communities, and their "possession" of knowledge might be analyzed in various ways, but there is usually no doubt that human knowers are the proper subject of knowledge. Even those philosophers who have previously emphasized the importance of practices for knowledge have typically located knowledge in the *agency* of knowers: practices are thereby understood to be what knowers do (or have the capacity to do) rather than just what they think and say. By contrast, my account of scientific knowing requires a more expansive notion of what a 'practice' is.[14] On my account, practices are not just agents' activities but also the configuration of the world within which those activities are significant. Attributions of knowledge are thus more like a characterization of the situation knowers find themselves within rather than a description of something they acquire, possess, perform, or exchange. The point is not to reject our ordinary ways of speaking about knowers. It can be perfectly appropriate to ascribe knowledge to a knower, so long as we understand that correct ascription of knowledge depends on how the knower is situated within ongoing practices rather than simply on whether the knower "possesses" the right beliefs or skills or stands in appropriate causal relations to facts.

14. A "deflationary" approach to understanding scientific knowledge is discussed in Chapter 7 below.

This understanding of scientific knowledge in terms of practices is not subject-centered. It therefore differs significantly from some of the best-developed philosophical discussions of practices, which in the end are reconceptions of human subjectivity. I find these "humanist" accounts of practices too readily compatible with more traditional conceptions of knowledge as something possessed or exercised by a knower. At least three distinct and influential humanist interpretations of the notion of a 'practice' exist. For Peter Winch and many others who have been similarly influenced by Wittgenstein, a practice consists of activity that is rule-governed.[15] Wilfred Sellars inaugurates and Robert Brandom further develops an alternative approach, for which social practices are the domain of justification and reason giving, which can neither be reduced to rules nor supplanted by causal explanations.[16] Hubert Dreyfus offers a third version, one that also rejects the primacy of rules but on the grounds of the irreducible role of embodied skills in practices.[17] Although they implement the distinctions quite differently, all three accounts of practices insist on both metaphysical and epistemological distinctions between the domain of practices and the domain of "nature" that is susceptible to causal explanation. My own account of practices is deeply indebted to both the Sellarsian tradition and Dreyfus but requires reconsideration of their underlying dualisms between nature and practices.

Rather than attempting a full exposition of Winch, Sellars and Brandom, or Dreyfus, I shall offer a direct account of the conception of practices which emerges from my discussion of scientific practices in *Knowledge and Power*. The important continuities and differences with their accounts (and with the social constructivist and materialist interpretations of practices within more recent science studies) will emerge in the discussion. My account of practices can be summarized in ten theses, each of which I will then discuss more extensively:

(1) practices are composed of temporally extended events or processes;

(2) practices are identifiable as patterns of ongoing engagement with the world, but these patterns exist only through their repetition or continuation;

(3) these patterns are sustained only through the establishment and enforcement of "norms";

15. Winch 1958: 25–39, 57–65, 83–86.

16. Sellars's own discussion of norms and social practices is best developed in Sellars 1963: chap. 4, and 1968: chap. 7. The clearest and most comprehensive discussions of a broadly Sellarsian notion of practices, however, are probably Brandom 1979 and 1994.

17. Dreyfus 1980, 1984, 1991, 1992.

(4) practices are therefore sustained only against resistance and difference and always engage relations of power;

(5) the constitutive role of resistance and difference is a further reason why the identity of a practice is never entirely fixed by its history and thus why its constitutive pattern cannot be conclusively fixed by a rule (practices are open to continual reinterpretation and semantic drift);

(6) practices matter (there is always something at issue and at stake in practices and in the conflicts over their ongoing reproduction and reinterpretation);

(7) agency and the agents (not necessarily limited to individual human beings) who participate in practices are both partially constituted by how that participation actually develops, and in this sense, 'practice' is a more basic category than 'subject' or 'agent';

(8) practices are not just patterns of action, but the meaningful configurations of the world within which actions can take place intelligibly, and thus practices incorporate the objects that they are enacted with and on and the settings in which they are enacted;

(9) practices are always simultaneously material and discursive;

(10) practices are spatiotemporally open, that is, they do not demarcate and cannot be confined within spatially or temporally *bounded* regions of the world.

With this overall sense of where the discussion of practices is going, we can now consider each thesis in turn.

First, the claim that practices are temporally extended is uncontroversial. Clearly the term 'practice' strongly suggests a recurring or otherwise continuous form of activity. Practices also clearly involve doings, although I initially describe practices more inclusively as events or processes rather than as actions. I opt for the wider category to leave open the possibility that practices incorporate the *setting* of action as well as the action itself. If there is anything controversial about this initial thesis, it will come from my subsequent claim that scientific knowledge must be located within practices. Knowledge has not usually been thought to be temporally extended in the way that practices clearly are: on most accounts of knowledge, whenever knowledge exists, that knowledge is fully present. Whether knowledge is identified with reliable or justified beliefs or with neurocomputationally described information, the knowledge itself is a state rather than a process.[18] The reasons or reliances that warrant it as

18. If knowledge is identified with appropriately warranted or reliable beliefs, these need not be occurrent mental states, but in whatever way nonoccurrent beliefs exist (for example, as a disposition), their existence is fully present temporally.

knowledge may be temporally extended, but knowledge itself is something fully present. My reasons for assimilating the temporality of knowledge to that of practices will become clearer in the next chapter, when I discuss the temporality of scientific practices as the ongoing narrative reconstruction of science.

Practices are constituted as patterns of events or processes, and the most obvious difference between the various humanist accounts of practices is their identification of what governs the continuation of these patterns: rules (Winch), communities' norms (Sellars/Brandom), or the purposiveness of skilled activity (Dreyfus). A significant dividing line is whether what governs the continuation of such patterns is identifiable apart from its actual instantiation, or whether the patterns exist only through their continuing reenactment. Winch here differs from Sellars and Brandom, Dreyfus, and me, because a rule could be formulated in the absence of any actual activity in conformity to the rule. For Brandom, by contrast, the communities that actually engage in a practice are the final arbiter of whether an event instantiates that practice: "The respect of similarity shared by correct gestures and distinguishing them from incorrect ones is just a *response* which the community whose practice the gesture is does or would make."[19] In the absence of the right kind of community, there simply is no question of conformity or nonconformity to a practice. Yet, in the end, what constitutes such a community on Brandom's account is a field of shared practices; there need be no objective criteria of correctness apart from the holistic regularities manifest in the community's ongoing activities and responses. Hence, in the absence of a sustained tradition of practice, no pattern would fix its identity.

Dreyfus offers different reasons for denying that the patterns that make up practices are identifiable apart from their actual instantiation in various circumstances. Skills do not exist except as the practiced repertoires of skilled agents. Furthermore, the skillful coping that is manifest in practices is a flexible responsiveness to situations; its purposiveness cannot be confined to predetermined formulas or already explicit purposes. Underlying Brandom's and Dreyfus's arguments, however, is a common theme: practices are open or flexible patterns of response which can be extended in novel ways and therefore cannot be confined to rules specifiable in advance. As shall become evident, my account of practices draws on both Brandom and Dreyfus: practices require skilled comportment and social norms.

19. Brandom 1979: 188.

Practices can, of course, die out when their constitutive pattern of activity is no longer reenacted or when the circumstances within which those activities would be intelligible no longer obtain. Skills can be lost, communities can dissolve, and social norms can be abandoned. Even then, however, the identity of practices is not restricted to their actual historical instances. Such closure is prevented by the fact that, at least in principle, even long-abandoned practices can be reactivated. This claim does not contradict the earlier claim that the patterns composing a practice do not exist apart from their actual instantiation. The requirement is for reinstitution of the kinds of communities and skills constitutive of practices and the appropriation of past practice as part of their constitutive pattern. Thus practices always include a horizonal future, as well as a history and an extended present.

A practice "includes" its future in a very strong sense: what the practice is at present to some extent depends on how its future develops. This identification of present practice with the past of a future that is not yet is heightened in the case of scientific practices, which are oriented toward the disclosure of what is presently unknown. As Hans-Jörg Rheinberger put this point in discussing the beginnings of modern cell biology and virology: "In the spontaneous recurrence of the scientist the new becomes something already present, albeit hidden, as *the* research goal from the beginning: a vanishing point, a teleological focus. Without the avian sarcoma virus of 1950 [Peyton] Rous' sarcoma agent would have remained something different. But: The virus of 1950 must be seen as the condition of possibility for looking at Rous' agent as that which it had *not* been: the *future virus*. The new is not the new at the beginning of its emergence."[20]

Next, the patterns of actions and events which make up practices are in a crucial sense normative. A pattern constitutes a practice rather than some other kind of regularity to the extent that it is a pattern of correct or appropriate performance. For Winch, an action is an instance of correct practice if it conforms to the rule constitutive of that practice. For Brandom, by contrast, instances of a practice are determined solely by the response of the community engaging in the practice. A performance is correct if the community responds to it as correct. Practices thus always encompass some events that count as differential responses to correct or incorrect continuations of the appropriate patterns, and the practice is sustained by the effectiveness

20. Rheinberger 1994: 77.

of these differential responses in creating and sustaining patterns of mostly correct continuations.[21] Dreyfus offers still a third account, for which success or failure in attaining the proximate end of the practice is constitutive of correct performance: for example, any activity that succeeds in driving nails is a correct instance of hammering. Dreyfus recognizes that norms are ultimately necessary for practices ("Without social norms there would be no way to set up the complicated referential totality involving nails, boards, houses, etc., in which driving in nails has its place, and so no role for [hammering]"),[22] but he insists that skillful practice is constituted more fundamentally by proximately successful performance. The norms that make sense of the ends of skillful activity are no more constitutive of a practice like hammering than are "the natural kinds and causal laws that make nails and pounding possible."[23] On all three views, of course, the "norms" that police and thereby maintain practices need not be explicitly articulated principles, nor do they necessarily function as justifications for the practices they sustain. The relevant norms are thus in the end "what is done," in the sense in which Wittgenstein suggested that justifications for a rule are eventually exhausted and Heidegger suggested that the "who" of everyday being in the world is *das Man,* "the anyone."[24]

My own view draws on both Brandom and Dreyfus. I reject Winch's view because it rules out the possibility of practices that change and develop over time. Violations of a rule that previously encompassed all and only correct instances of the practice may nevertheless mark the continuation of a significant pattern of correct performance. But Brandom and Dreyfus also make a parallel mistake. Each proposes a *final* arbiter for the norms of correct performance of a practice: Brandom concludes that how the relevant community responds is constitutive of practical norms, whereas Dreyfus appeals to the implicit telos of skillful activity. Moreover, the finality of these criteria stems from their significance for Brandom's and Dreyfus's understanding of

21. As Brandom astutely notes, however, such differential responses to correct and incorrect practice often count as such normative responses only on the grounds of further differential responses to them. The result is not an infinite regress but an interconnected web of practices identifiable only by interpretive participation or translation. Although the same behavior could also be accounted for causally, the claim is that such a causal accounting would overlook significant patterns within the same complex of events. Dennett (1991b) argues for the reality of such patterns that require the right kind of interpretive stance in order to be discernable.

22. Dreyfus 1992: ms. 4.

23. Dreyfus 1992: ms. 4.

24. Wittgenstein 1953: par. 217; Heidegger 1962: sec. 25–27.

what it is to be a free, human subject: a recognized member of a community or a skillful body. Brandom and Dreyfus allow that human beings become subjects only through their participation in practices, but they do not allow that what it is to be a subject is likewise situated historically and culturally. In any case, neither proposed criterion can have the alleged constitutive finality, because neither who belongs to the community nor what would count as a significant end for assessing the success of skilled practice can be fixed in advance.

In Brandom's case, who belongs to the appropriate community is not and cannot be settled in advance of the ongoing patterns of performance and response. What has heretofore been a community can fracture over precisely the ends for the sake of which they enforced norms of correct practice. Moreover, the fracture can itself be contested: some practitioners will still see themselves as belonging to one community, whereas others will insist that there are at least two, or none at all. Dreyfus, however, also cannot (as he would like) dismiss communities and their norms as necessary but not constitutive for practices, because skillful practice is ultimately determined by the ends of a practice (and hence by the social norms that determine the appropriateness of skills). A presumptively skilled practice like hammering counts as skilled only if the patterned transformation of the world that those skills bring about counts as a significant achievement. Yet the significance of the achievement depends on the ways in which others respond to it. As we shall see, the patterns constitutive of practices require ongoing adjustments of practitioners' relations to those they work with, to what they work on, and to whom and what they work for; none of these relations automatically takes priority in sustaining an ongoing pattern of practice.

The fourth thesis indicates that practices continue to be enacted only to the extent that their continuity is sustained and enforced by normative sanction, that is, by differential environmental responses to actions which sustain or reinforce the practice and those which relevantly oppose or deviate from it.[25] We can accept Brandom's claim that an event counts as correct or deviant practice on the basis of

25. It will not escape the attentive reader that, at least in this rudimentary way, I count as practices the various behavioral patterns of organisms, thus clearly not confining the notion of practice to the human sciences. The patterns of activity that compose practices are identifiable only by a "normative stance" closely parallel to the "intentional stance" that Dennett (1987) sees as the enabling condition for the manifestation of the patterns that make up intentionality. Okrent (1991, forthcoming) offers an account of intentionality which emphasizes teleology and agency and which explicitly incorporates the more rudimentary intentionality of nonhuman organisms.

how those recognized as practitioners respond to it (although their response may be crucially mediated by the effect of the activity on other practitioners or things). Moreover, those differential responses must in some way be relevant to whether and how the practice is continued: responses to correct and incorrect practice must prima facie count as encouraging or discouraging the continuation of correct or incorrect practice, respectively. But the responses need only be prima facie effective (rather than, say, causally determinative) for a reason that Brandom articulates in a slightly different context. For him, the difference between freedom and causation, between practices governed by norms and movements governed by causal law, must be understood to be a social and practical rather than an objective difference.[26] Practices depend not on the objective freedom of practitioners but on whether practitioners understand and respond to one another *as* capable of acting in accordance with norms. Similarly, the issue raised by the differential environmental responses that demarcate correct from incorrect practice is not whether those responses are causally efficacious in sustaining patterns of correct practice, but only whether those responses continue to be significant (that is, whether there continues to be a discernable pattern of practice within which those responses that differentiate correct and incorrect practice *count* as making a difference to continuing practice).

It is therefore central to the notion of a practice that practices are understood by their practitioners *as* enforced. This claim in turn implies that practices can exist only against a background of resistance.[27] New practitioners must be socialized into practices, deviations by their predecessors must be deflected, suppressed, or contained, and the environment must be (re-)arranged in ways conducive to their ongoing reenactment. Thus, practices always include (and to some extent, constitute) power relations. Indeed, as we shall soon see, there are close and important connections between power relations and the practices that embody scientific knowledge.[28] These connections are often misunderstood or overlooked, however, not only because of the reification of scientific knowledge but also because many

26. Brandom 1979: 190–92.

27. As will become clear in subsequent chapters, 'resistance' encompasses both the recalcitrance of things and the nonconformity of agents. This usage is not a conflation of distinct kinds of phenomena but a recognition of their inextricable intertwining and interchangeability.

28. The relations between power and knowledge will be more extensively discussed in subsequent chapters, especially Chapters 7–9.

of the standard interpretations of power within the literature of social and political theory are inadequate for understanding the place of power relations within the practices where scientific knowledge is located.[29]

These relations of difference, power, and resistance which are partially constitutive of practices cannot simply be characterized as struggles between the defenders and the opponents of particular practices. My fifth thesis shows why: such a description is inappropriate because it mistakenly suggests that the identity of a practice (as well as those of the interests and groups who support or oppose it) has already been fixed by its history and its ongoing reenactments. Taking practices to be fixed in this way underestimates the importance of their normativity and their lack of temporal closure. Instead, the very identity and significance of a practice is ordinarily at issue in the conflicts and differences over its ongoing reenactment. As a result, practices are radically open: whether a subsequent action counts as a continuation, transformation, deviation, or opposition to a practice is never fixed by its past instances.[30] These instances are, of course, relevant to the identification and continuation of a practice, but they cannot be decisive in settling whether new cases exemplify the practice; the new cases themselves may, after all, constitute a reinterpretation of their predecessors. Social constructivist interpretations of practices fail to take adequate account of this openness of the social dimensions of practices. When they insist that social relations or interests are explanatory, they foreclose the possibility that those relations or interests, or even their characterization as social, may be what is at issue in the continuation of the practice.

This radical openness of practices can be aptly characterized by appropriating Alasdair MacIntyre's description of the openness of historical traditions: "What constitutes a [practice] is a conflict of interpretations of that [practice], a conflict which itself has a history susceptible of rival interpretations."[31] These conflicting interpreta-

29. For a useful critical discussion of some of the most influential literature on power (along with a more adequate, constructive account of power which both complements my discussion in Rouse 1987b and anticipates many important themes from the discussion below), see Wartenburg 1990.

30. Kripke (1982) has prominently argued for the comparably open (or practice-dependent) character of rules. Although Kripke's discussion raises interesting parallels to my discussion of practices, a crucial difference is that he (mistakenly, in my view) regards the open-endedness of rules as raising distinctively skeptical problems about whether an action could ever count as following a rule.

31. MacIntyre 1980: 62.

tions of a practice need not be explicitly articulated, however. One also interprets a practice simply by participating in it; one's participation reinterprets what the practice is even as it thereby continues the practice. Practices are therefore open to semantic drift as well as principled reinterpretation (and the distinction between the two cannot be a sharp one, because the relevant principles are not themselves protected from drifting in turn). As Steve Fuller argued about linguistic practices specifically: "These practices exist only by being reproduced from context to context, which occurs only by the continual adaptation of knowledge to social circumstances. However, there are few systemic checks for the mutual coherence of the various local adaptations. If there are no objections, then everything is presumed to be fine."[32] Under such conditions, of course, considerable transformation and diversification of a practice can take place more or less unnoticed. Fuller's remark thus reemphasizes the significance of norms in shaping practices: the degree of continuity in the ongoing reproduction of a practice depends importantly on the extent to which such continuity is enforced by normative response. The less thoroughly and rigorously the practice is policed to discourage "deviation," the more susceptible it is to drift. But such policing is itself no guarantor of continuity in a practice; the normative responses that police practices are themselves practices, similarly subject to semantic drift.

Part of the difficulty in specifying a practice by its constitutive patterns is that practices are defined not only by the specific activities that compose them but also by what those activities are about (what is "at issue" in the practices) and by what is at stake in their success and their continuation. As my sixth thesis indicates, practices matter, and how and why they matter and to whom are crucial to whatever integrity they have as practices. There is an important sense in which one has not understood a practice unless one has grasped the point of the practice, that is, what is at issue and what is at stake. The recognition that practices are focused by such issues and stakes does not, however, challenge my earlier insistence on the openness of practice. The issues and stakes in practices are not themselves protected from reinterpretation and drift any more than are the ongoing patterns of activity which take up those issues for those stakes. If anything, the reverse is true: precisely because what is at issue in a practice and what is at stake in conflicts over those issues are assumed to be what "everybody

32. Fuller 1989: 4.

knows," these issues and stakes often remain only partially articulated and therefore ironically less susceptible to normative constraint. Indeed, on many occasions, the very attempt to articulate how a practice matters signifies a recognition that considerable drift has already taken place and thus a nostalgia over the loss of a presumed prior unity of purpose.

My seventh thesis, concerning the role of agents in practices, requires more extended discussion. The various humanist interpretations of 'practice' focus on what seems to be an obvious and crucial difference between practices and other recurring patterns within the world, namely, that practices are always done by someone or other. Moreover, something is crucially at stake in this point: agents, that is, those who engage in practices, may have moral and epistemic obligations to one another which differ significantly from their appropriate relations to nonagents. For example, Brandom is concerned with the Kantian conception of free beings whose rationality and dignity call for respect, whereas Brandom, Dreyfus, and Winch all insist that agents' practices are to be translated or interpreted rather than explained causally. I agree on the importance of this point and its consequences but construe the issue differently. On my account, practices always involve doings and doers, along with what these doings are done with and done on. Practices are thus not just what agents do but also the relational complex within which their doings are intelligible.[33] The agents who engage in practices thus belong to the practice, rather than the reverse.[34] Agents' belonging to practices is complicated, however. Just as the objects utilized within a practice are usually not fully constituted by that practice (the same objects may also show up in a variety of other practices), so the same agents who engage in one practice also typically participate in many others. In this sense, agents obviously do exist outside the practices in which they participate and are not simply part of any of those practices. Yet agents do not have an identity fully separate from the practices in which they participate, such that who or what the agents are is partially constituted and/or transformed by those practices. Agents therefore "belong" to practices in both a weaker and a stronger sense. On the one hand, an

33. This formulation and the need for it emerged in my conversations with Mark Okrent, whom I gratefully acknowledge.

34. The agents who engage in practices need not, of course, be individuals. Indeed, who or what can count as an agent is itself at issue within various practices and is not established ahistorically by a timeless nature of agency or of the various kinds of putative candidates for agency.

adequate description of a practice necessarily includes a characterization of the agents who (might) participate in it. In the stronger sense, an agent belongs to a practice to the extent that an adequate understanding of the agent (and the agency of the agent) includes an account of its participation in that practice.

The significance of this point turns on the basis for distinguishing agents from nonagents. For Dreyfus and Winch (on different grounds), there seems to be an objective difference between those events that belong to the domain of the natural sciences and those that cannot be adequately understood unless taken *as* skillful comportment or rule-governed action. Brandom, by contrast, denies an objective difference between translatable practices and explainable actions, for he regards the same events to be understandable in either way (albeit with greater difficulty in some cases than in others). The difference between agents and nonagents is instead a social and practical difference, but with comparable moral and epistemic consequences. For Brandom, that is, who is a rational agent, entitled to respect and worthy of translation, is determined by whom a community recognizes as such. My account of practices follows Brandom in denying any objective difference between agents and nonagents; the differences are instead constituted within practices and are intelligible only in terms of discriminations that get their identity and significance from practices. Nevertheless, community judgments are not fully authoritative in the way Brandom proposes.

The difficulty once again is that the identity and boundaries of communities need not be clearly or unambiguously specified. Who counts as a member of the community of agents or persons is supposed to be determined by the responses of members of that community to one another (for example, which events are taken as rational actions to be understood or translated and which are taken as natural events to be explained or manipulated). These differences in types of response are themselves differentiated in the same way; what counts as appropriately responding to someone as an agent is adjudicated by the community's response, and infinite regresses are halted by holistic regularities of performance. Brandom concludes that "for us to engage in a web of social practices no such requirement [of an objective criterion of correctness] applies; all that is required is sufficient agreement within the community about what counts as an appropriate performance of each of the practices."[35]

35. Brandom 1979: 191.

Yet such agreement that is sufficient to enable practices to proceed without infinite regress may nevertheless still involve resistance and dissent. As Samuel Wheeler noted, words like "community" or " 'culture' hide a diversity of incompatible groups, . . . a loose assortment of coexisting, overlapping, and interacting groups and individuals."[36] Even though not everyone participates equally in determining what counts as community responses, the dominant interpretations are nevertheless not fully authoritative everywhere. Thus, the boundaries of the community of free and responsible agents have often been contested and ambiguous: differences of gender, race or nationality, social status or class, and so forth have frequently marked out these contested regions, within which " 'we's that are not heard with authority, who do not get to decide what gets said when, still get to be heard, sort of. . . . That 'culture' is an oversimplification does not necessarily mean that no interests are in charge, just that the charge cannot be total."[37] The ambiguity results from the interconnectedness of practices. Attempts to maintain the social dominance of one or more groups may require adjustments elsewhere in what they say and do, or what they recognize as acceptable or intelligible sayings or doings, and these adjustments may not work out fully coherently. We cannot say in advance where these adjustments will break down, and hence Brandom is not quite right that "the community has final say over who its own members are."[38] As Wheeler trenchantly notes:

> A hegemonic discourse, typically, cannot keep people in line [because] we have something like reality interfering with a construction imposed by the dominant. The dominant ideology [or] cultural "theory" is an organization of pleasure and pain, of what is valuable and despicable. But the "data" in this case, though they are not independent of the hegemonic conceptual scheme, yet resist it. Things just are not working out very well for the underclasses. Pleasure and pain, in practices, are something like "observation" in science. There is no "pure data" but the world troubles "theory" nevertheless.[39]

If the difference between agents and nonagents is established within practices, then, this difference is determined neither objectively nor

36. Wheeler 1991: 208.
37. Wheeler 1991: 209–10.
38. Brandom 1979: 191.
39. Wheeler 1991: 209.

socially. What matters is interconnected and ongoing practical interaction with the world.

This point about the place of agency within practices is very important for understanding scientific practices. Social constructivist accounts of scientific practice or agency have typically followed Brandom in insisting that in matters of social practice, the authority of communities over norms of correct or incorrect performance is unconstrained (social constructivists, of course, crucially diverge from Brandom by including natural scientific knowledge among matters of social practice; they accept the terms of his dichotomy between social and objective thing-kinds, while insisting that the latter class is empty).[40] In my account of scientific practices I deny that their outcome can be accounted for either as objectively determined (whether by real natures of things or rational necessities of inquiry) or as socially determined. The normative configurations of the world which emerge in practices must be understood as prior to any distinctions between agents and their environment, the social and the natural, or the human and the nonhuman.

This recognition that the contested boundaries between agents and nonagents cannot be resolved either objectively or socially has especially important consequences for understanding the political and cultural significance of science and technology. The boundaries between those beings that have moral standing worthy of respect and those beings that may be regarded as instrumental to the ends of morally significant beings are frequently at issue in controversies over scientific and technological practices. The agency and moral dignity of ecosystems, endangered species, or the biological environment ("Nature") have been fundamental questions for environmentalist politics and science. Many feminist theorists have found that gender politics are not simply about relations among men and women but are focused precisely on how to understand agency, body, rationality, and the boundaries between nature and culture, and such theorists thus regard "the science question" as central to feminism. Questions about the moral standing of human research subjects, laboratory animals, fetal tissue, or genomes have been addressed directly at research practices themselves. Humanist and social constructivist

40. Many of the classic social constructivist texts (for example, Bloor 1991, Barnes 1982, Pickering 1984, Collins 1992, Woolgar 1988c) formulate this claim in a Winchean, Wittgensteinian, or Quinean idiom of rules, conventions, forms of life, networks, or webs of belief, rather than following Brandom's appeal to social practices. Mutatis mutandis, the point is the same.

conceptions of practices are not well situated to understand these dimensions of contemporary scientific and political understanding because they offer conceptions of agency and knowledge which would settle the controversies in advance, by philosophical fiat.[41]

I can now consider my eighth thesis. As the preceding discussions should have clearly indicated, the various humanist accounts of practices can be understood as attempts to reinterpret intentionality.[42] Rule-governed behavior, responsive social recognition, or skillful comportment are alternatives to mental representation as a characterization of meaningful directedness toward the world. Yet the humanist interpretations of practices still betray an important continuity with the Cartesian tradition they aim to displace. The classically Cartesian conception of mental representation takes thoughts "in" the mind to be ontologically distinct from objects "outside." Whether mental representations are substantively distinct from things or belong to a distinct level of description of things that could also be described physically, their "subjective" character is significantly different in kind from the objective character of physical objects. Moreover, philosophers who emphasize mental representation have attempted to specify representational content as "narrowly" as possible, that is, they aim wherever possible to use only relations among mental representations, without reference to the natures of or causal interactions with the objects purportedly represented.[43] Practices on any account seem rather less opaque to the world than are mental representations in this sense. Yet the humanist accounts of practices still parse them into "inside" and "outside," whether the "inside" of a practice is its constitutive rule, pattern of social recognition, or skillful engagement. The practice is the subject's mode of directedness beyond itself.

To understand my divergence from such humanist accounts of practices, it may be useful to take quick stock of the argumentative terrain over intentionality. Appeals to practices are among the many

41. I have discussed the political significance of scientific interpretations of nature at greater length elsewhere, notably Rouse 1987b: chaps. 6–7, and 1991b. Haraway (1989, 1990, 1992a,b); Latour (1987, 1993); and Pickering (1995) have also insisted in different ways that science studies cannot take for granted any specific boundary between natural and social worlds or between natural objects and human agents.

42. Thus, for example, Brandom's claim that no objective difference exists between actions according to norms and causally explicable behavior is closely akin to Daniel Dennett's claim that intentionality is not a property of things but an interpretive strategy whose appropriate application to any particular system depends on the purposes of the interpretation. See Dennett 1987.

43. For a contemporary programmatic account of such a Cartesian conception of mental representation, see Fodor 1980.

philosophical alternatives to naturalistic accounts of meaning as fully explicable causal processes of information flow or naturally selected responses to an environment.[44] Nonnaturalistic accounts typically see meaning as *bestowed* rather than thus found among the natural constituents of the world. Nonnaturalists classically locate the bestowal of meaning in the "intrinsic" intentionality of the mental,[45] but the humanist accounts of practices locate intentionality in the irreducible interaction of subjects with one another and their surroundings. In the most prominent versions, this interactive "bestowal" of meaning is accomplished either by interpretation (Davidson or Dennett), social recognition (Brandom), or skillful comportment (Dreyfus's interpretation of Heidegger). I take something right to be captured by each approach, but all three also err in still opposing a vestigially subjective "bestowal" of meaning to its naturalistic emergence.

Brandom helpfully focuses the issue when he proposes his alternative way of thinking about traditional distinctions between subjects and objects: "Social practices constitute a thing-kind, individuated by communal responses, whose instances are whatever some community takes them to be. *Objective* kinds are those whose instances are what they are regardless of what any particular community takes them to be."[46] The distinction is still between kinds, but "subjects" are reconstrued as social patterns of recognitional response rather than individual minds or mindings. In effect, Brandom distinguishes the "social world," one in which community responses are authoritative, from the "natural world" of objects whose determinations "are what they are regardless of what any particular community takes them to be."[47] He thereby tries to stake out a middle ground between naturalists, for whom the meaningfulness of the "social world" is naturally explicable, and social constructivists, who would insist on the ontological priority of the social world through which community responses do determine what is (that is, what can count as) a natural determination. I attempt to reject all three positions by rejecting the prior distinction between autonomous social and/or natural "worlds." Social constructivists are right to recognize that "natural" objects do not

44. Influential contemporary versions of biological and information-theoretic accounts of intentionality are, respectively, Millikan 1984 and Dretske 1981.

45. For a sophisticated contemporary discussion of the intrinsic intentionality of the mental, see Searle 1983.

46. Brandom 1979: 188.

47. Brandom 1979: 188.

"have" identities or determinate properties and relations outside a field of practices,[48] but are mistaken (along with Brandom) in taking community responses to be authoritative within that field. As we have already seen, who belongs to the relevant community may be what is at issue in further articulations of a practice and not what determines their outcome. Part of the error is in underestimating the significance of the materiality or "embodiment" of social practices, as well as overestimating the clarity and unity of "communities" and their responses to new performances. An appropriate remedy is to understand practices as the field within which both the determinations of objects and the doings and respondings of agents emerge as intelligible. Such a conception is still nonnaturalistic (the normativity of practices cannot be reduced to causal relations), yet it eliminates the vestigial sense that meaning is something *bestowed* by the activity of subjects. This latter point can be indicated in two distinct ways. First, as I suggested above, who or what can count as subjects or agents, along with what it is to be an agent, is historically situated; it emerges from a field of practices rather than constituting it. Second, the ongoing patterns that compose practices are patterns of the world in which agents' activities are situated. One might say that the role of active subjects in configuring practices is not the bestowal or imposition of meaning but a responsiveness to the solicitations of their situation.

This claim undoubtedly seems perplexing initially, but Dreyfus's attempt to found intentionality on skillful comportment may explicate it. The contribution of Dreyfus's account is twofold, both in emphasizing the bodily aspects of practices and in recognizing that the identity of practices depends on the right kind of situation or environment as well as on the right kind of activity. On the first point, Brandom's account of social practices is by contrast oddly reminiscent of functionalism in the philosophy of mind in its seeming indifference to how social practices are materially realized. Or perhaps more salient, it is reminiscent of theories of language as an ideal structure whose particular phonetic, ideographic, or alphabetic materializations are of no significant import. For Dreyfus, the paradigm cases of intentional directedness are neither linguistic nor perceptual representations but skilled bodily activities. Such activities are gradually

48. For a detailed discussion of this rejection of realist accounts of the mind- and language-independence of the "natural" world and how it is compatible with the commonsense recognition that things are not always (or often) what we try to make of them, see Rouse 1987b: chap. 5.

emerging repertoires of bodily possibilities and clearly draw on the material contours of agents' bodies.[49]

The second contribution of Dreyfus's account is perhaps more crucial, however. Skilled comportment is not an imposition of meaning on things but a meaningful responsiveness to them. To be sure, skills engage things under aspects: deftly drawing with a pen responds to it differently from accurately throwing it, and its relevant "properties" depend on these particular purposes. Yet skillful use of things is always responsive to their particular capacities; someone who knows how to write readily adapts to the distinct characteristics of ballpoint, felt, and fountain pens. Furthermore, one cannot engage in skillful activity without the right sort of equipment in the right surroundings. Deliberately attempting the "same" movements in the absence of an appropriate object requires both a physically different movement (one's bodily comportment in such practices responds to the particular resistances actually offered by the object)[50] and constitutes a significantly different activity (miming or pretending). In skilled activity as a mode of intentional directedness, the relational complex of the agent situated in a particular setting is in an important sense prior to both the agent's actions and purposes and any independently definable properties of surrounding things. Things become manifest according to their significance for possible action, while the agent's activities and possibilities are configured by what her surroundings afford.

I am characterizing practices more generally as situated patterns of activity. I thus want to extend this priority of the situation (the relational complex of embodied agents in meaningfully configured settings for possible action) over the subject/agents and objects they incorporate. Dreyfus interprets more complicated intentional performances differently: he accepts more traditional representationalist accounts of such performances but regards them as somehow "founded" on the "original" intentionality of comparatively simple cases of skilled comportment. I think we are better off without such a pragmatic foundationalism, both because the simple cases only ac-

49. Butler (1987) disentangles the materiality of bodies from any notion of the body as "natural," describing the body as a cultural situation to undercut distinctions between the cultural interpretations of gender and the natural reality of sex. Butler's specific discussions of sex and gender extend to Dreyfus's conception of skilled comportment, so that the intentionality of skilled performance can be thoroughly material, without presupposing a natural substratum of already-given bodily possibilities.

50. This characteristic of our everyday practical dealings with things has been noted by Merleau-Ponty 1962: pt. 1, chap. 3, and emphasized by Todes 1966.

quire their significance as skillful practice within larger configurations of practice (see the discussion of the third thesis, above) and because Dreyfus's desire to treat social and linguistic practices differently seems neither well motivated nor adequately justified. Admittedly, the dependence of the supposedly "nonoriginary" practices on appropriate material settings is more complicated because most interesting practices involve multiple relations to objects in the world, which can function as equipment, proximate objects, obstacles, surrounding environment, and much more. Practices also involve complex social relations and norms in more obvious ways. Yet this point just reminds us of the complexity of any analysis of the situations that we negotiate in practice with ease.

Scientific practices are an especially instructive case because they clearly involve the kinds of skillful manipulation and discrimination that Dreyfus would characterize as "originary" skilled performance.[51] Yet any attempt to disentangle this kind of skilled comportment from the social norms and linguistic discriminations that are supposedly "founded" on individual coping skills seems unpromising. Indeed, it suggests difficulties reminiscent of those confronting more traditional foundationalisms that needed to disentangle foundational sense experiences from the linguistic and/or theoretical constructions through which they are interpreted. Moreover, scientific practices seem especially clear cases of activities that cannot be sensibly understood apart from their local material settings; that was the central point of my extended discussion in *Knowledge and Power* of the construction and manipulation of controlled microworlds within the institutional settings of laboratories, clinics, or field sites.[52]

Thus, despite his appropriate and important recognition of ways that practices are embodied and situated, Dreyfus's attempt to construe intentionality as founded on skilled comportment has three closely connected inadequacies. The first of these inadequacies I have already discussed elsewhere, namely, Dreyfus's explicit essentialism about natural science and his correlated attempt to preserve an attenuated realism about the natural kinds and causal relations that he

51. Experimental and instrumental skills are obviously prominent here, but Dreyfus's critical studies of artificial intelligence research have also led him, rightfully, to include intellectual capabilities such as chess playing among his originary skilled bodily comportments. Presumably, skilled scientists can "inhabit" theoretical models and respond to their solicitations in specific problem situations in much the way Dreyfus claims grand masters can size up the situation on the chessboard.

52. The argument for the significance of local knowledge of controlled microworlds is in Rouse 1987b: chap. 4.

takes the practices of "modern science" to reveal.[53] Second, I believe Dreyfus is also mistaken in his attempt to construe skilled comportment as an asocial basis for intentionality. As I noted above, he is unable to account for the normative determination of skillfulness without referring to social responses to putative skills and achievements. Moreover, I suspect that there is at least one significant connection between these two points: despite his insistence that bodily skillfulness cannot be understood naturalistically, the presumed causal individuation of bodies might strongly suggest that bodily intentionality must be at bottom individual.[54] Finally, the asociality of Dreyfus's account is explicitly connected to his attempt (mistaken on my view) to identify skilled comportment as somehow prediscursive. Dreyfus insists instead that social normativity "only becomes constitutive at the linguistic level, and language, itself, is built upon the significance-structure revealed by skilled activity."[55] I discuss this point below, when I take up more generally the question of the discursiveness of practices.

But before turning to discursive practice, it may be useful to recapitulate the point of this long and involved discussion of my eighth thesis. I had argued above that practices should be understood not as a surrogate conception of human subjectivity but as the larger relational complex in which subjects (agents) and subjectivity (agency) are constituted. I follow up that argument here by considering how practices might be understood to incorporate the situations within which actions are intelligible. The central claim is that the meaningfulness of agents' situation is not bestowed by either individual comportment or by social norms in abstraction from their material realization; instead, it emerges from the ongoing interaction of agents with their material surroundings and with one another. The situation in which agents find themselves is already meaningful, not because meaning is grounded in natural causality but because agents are always responding to the specific configuration of meaningful possibilities for action which emerges from past practice. The temporality of

53. See Rouse 1987b: chap. 6, and Rouse 1991b. What Dreyfus is committed to that I reject is the essential unity of the natural sciences as "decontextualizing" practices and the correlated unity of the causal structure of the world as manifested decontextually. For a more thorough criticism of these specific commitments about the sciences (without specific reference to Dreyfus's adherence to them), see Dupre 1993.

54. The growing feminist literature on considerations of nature and culture in the individuation and construction of bodies is a helpful counterpoint to the seemingly natural unity and identity of individual human bodies. See especially Butler 1990, 1993.

55. Dreyfus 1992: ms. 8.

practice thus becomes the basis for an alternative to both naturalistic and subjectivist accounts of intentionality. The importance of the temporality of scientific practices will be the specific focus of the next chapter.

As for my ninth thesis, the question of the place of discourse or signification within a larger conception of practices and the linked question of how to understand discursive practice require far more extensive discussion than I can provide here. My limited goal is to indicate how these questions and my responses to them fit into the larger conception of practices which I am outlining here. A more extended discussion of how I would address this topic for the purposes of science studies can be found in Chapter 8 below. I argue against two quite distinct approaches or tendencies. First, I want to oppose representationalist accounts of language and of the significance of language in scientific practices and scientific knowledge. That is, I want to understand language and language use in terms of discursive practices rather than as meaningful representations of states of affairs.[56] Second, I oppose attempts to understand practices in ways that subordinate or diminish their discursive dimensions. As I noted above, Dreyfus's account of practical intentionality is one such case. Influenced by early Heidegger and by Merleau-Ponty, Dreyfus understands skillful practice as a bodily engagement with the world which is in some sense more "fundamental" than language (partly, I suspect, because in the end, language is being construed representationally). A more crucially relevant case, however, is what I described above as materialist interpretations of scientific practices. I want to oppose a philosophy, sociology, or historiography of science which would reduce the linguistic dimensions of scientific practice to something like "rhetorical force" or "literary technology."[57] Signification in scientific practice (including metaphors and models as well as supposedly "literal" discourse) is too rich, inventive, and important to be adequately understood in these terms.

The two tendencies I am opposing have an adversarial relation. An important motivation for more recent emphasis on the materiality of social practice within science studies has clearly been dissatisfaction

56. For discussions of representationalist and nonrepresentationalist accounts of language, see Brandom 1976 and Rorty 1991.

57. What I am calling the materialist interpretation of scientific practices is more a tendency or theme than an explicit theory of scientific language. Prominent examples of work that tends in this direction are, among others, Shapin and Schaffer 1985; Smith and Wise 1989; Fuller 1988, 1989; Latour 1987; and Galison 1987.

with the tradition in philosophy and internalist history of science which largely presumes representationalist accounts of scientific language. A comparable motivation for work in the internalist and representationalist tradition has been to foreground the discursive dimensions of scientific practice. The core of my response is that one can emphasize the materiality of scientific practice consonant with the work of the "practice industry"[58] while still doing justice to the discursive richness of scientific practices. The crucial move is away from conceptions of signification as ideal, that is, as material expressions of a meaning or content that could be identically expressed in other ways.[59] Frege's and Husserl's classical accounts were rejections of naturalism in semantics; my response is not a return to doctrines of natural signification but the claim that signification is always an irreducible configuration of the world itself. We can begin to see this point by reflection on the ostensibly anti-Fregean claims of scientific realism. As I argued earlier in the book, realists need to distinguish the way the world is from the way our representations describe it as being, in order to ask whether and how the one can come to correspond to the other. My approach rejects the realists' understanding of both sides of this supposed contrast. The world is not in one "way" rather than another apart from the signifying practices through which possible alternatives can be discriminated and articulated, but these practices signify only within a world (so that they count as signifying and as signifying in some particular and meaningful way, only through their regular, ongoing connections to a surrounding world).

Another way to indicate this claim is that there is no way out of, or "behind," language, a point that Quine and Davidson tried to capture by substituting practices of translation or interpretation for the "idea idea" while insisting that such interpretation always takes place against the unanalyzed background of a home language. There is no fixed reference point where signification and interpretation can be anchored, neither in the natural segmentation of real kinds of objects nor in the structure of a language, conceptual scheme, social context, or culture. Interpretation always comes to an end in practice (action is not ordinarily disabled by indeterminacy), but for pragmatic reasons and not because it has reached a final resting place.

Not all material things signify, of course. To that extent, it may

58. I owe this expression for the materialist tendencies in science studies to Betty Smocovitis.

59. The classical accounts of the ideality of signification are Frege 1970 and Husserl 1970: Investigation One.

seem as if signification is located within a specifiable part of the world, namely, those things that function as signs. Yet signs signify only by belonging to more wide-ranging practices that open up to the world as a whole. Only by reference to the surrounding environment as configured by ongoing practices (including signifying practices) do signs count as more than marks and noises,[60] yet those practices themselves are what enable the world to display itself as having a significant configuration. This point is difficult to grasp. I felt that I first had a handle on it when reflecting upon Wittgenstein's discussion of "expecting": "It is in language that an expectation and its fulfillment make contact."[61] Wittgenstein notes that the various behaviors and surroundings that count as "expectings" do not have some unifying properties (or even family resemblances) prior to the practice of using the term 'expect' to talk about them. To that extent, the linguistic practice introduces this possible discrimination into the world. Yet 'expect' does not already have a meaning apart from the ongoing practices of using it in such situations; the signification and the significant configuration of the world happen together (and both endure only through the ongoing reproduction of the practice). When we adequately understand this characteristic of signs, we can see that they are thoroughly material; they signify through their *actual* repetition as the same under different circumstances, not through any ideal replicability. Their signification is thereby located within the world without its location having clear bounds. The openness of signifying practices is thus spatial as well as temporal. They open on the world as a whole even as they are locally situated within it.

As the final thesis, the claim that practices are materially located within the world without being identical to any temporally or spatially bounded region of it is very important for understanding scientific knowledge as situated within practices. It may be useful to understand this claim as an alternative to eliminativist interpretations of naturalized epistemology which replace talk about knowledge and justification or reliability with discussions of the causal structure of some clearly bounded region of space-time (for example, the neurophysiological eliminativism that considers causal processes within the

60. This point should not be taken to indicate that marks and noises can be disentangled from their semantic significance, any more than natural bodies can be separated from their culturally interpreted possibilities (see n. 49 above).

61. Wittgenstein 1953: par. 445. The discussion of 'expecting' encompasses pars. 444–53 and 569–86; I am grateful to Meredith Williams for first bringing the importance of these passages to my attention.

organism or its central nervous system, or the sociological eliminativism that examines specific networks of social causation).[62] The eliminativist is able to eliminate all talk of knowledge (and related epistemic terms such as truth, justification, reasons, and so on) because the location of knowledge within the world is bounded: knowledge can be equated with an (in principle) fully specifiable domain of causal processes, whether those processes are located "in the head" or in the interactions of social agents. But this claim that knowledge is spatiotemporally bounded conflicts with the spatiotemporal openness of practices, such that if knowledge is located in practices, it cannot be spatiotemporally bounded.

The reason that this conflict emerges goes back to my earlier discussion of the normative and open-ended character of practices. Practices are iterable, and their iterations are always subject both to deliberate reinterpretation and to semantic drift. Even if scientific inquiry as it has developed historically as a knowledge-producing practice were adequately described by an eliminativist interpretation (for example, as the causal interaction between the central nervous systems of individual organisms and the physical information in their environment, or as the causal product of complex networks of social exchange), this fact would provide no guarantee that the subsequent historical development of the same epistemic practices would continue to be so describable.[63] One way of seeing this point relies on the normative character of practices: any such confinement of epistemic practices within spatiotemporal bounds would require that such confinement be enforced, and the practices that enforce it need not themselves be confinable within the same spatiotemporal boundaries.

This claim suggests a more general lesson, however. Practices are always interconnected, never existing in isolation from one another, in ways that fundamentally affect the ongoing development of any particular practice.[64] A typical practice needs other practices to en-

62. I further discuss the differences between my deflationary conception of knowing and epistemological eliminativism in Chapter 7 below.

63. This claim by itself does not provide an argument against any particular version of eliminativism but serves only to differentiate more clearly the claim that knowledge is located in ongoing practices from any eliminativist interpretation of knowledge. Eliminativists, after all, identify knowledge (and/or learning) with a specific domain of causal processes as itself a matter of natural (or social) necessity. Hence, the eliminativist wants to rule out in advance the possibility that practices of inquiry could change in so fundamental a way as to be no longer identifiable with the relevant domain of causal determination.

64. Although his account is not articulated in terms of practices, in his discussion of "being-in-the-world" and the "worldhood of the world," Heidegger (1962: esp. secs. 12–27) works out a similar interpretation of human comportment, emphasizing its interconnection as a significant whole.

force its norms, provide its necessary equipment and resources, educate and train its practitioners, confer significance on it or undercut its previous significance, conflict with its continued development in any of the foregoing ways, and, in general, help configure the world in ways that allow the practice to be intelligible. In order to understand how scientific knowledge is situated within practices, we need to take account of how practices are connected to one another, for knowledge will be established only through these interconnections. Scientific knowing is not located in some privileged type of practice, whether it be experimental manipulation, theoretical modeling, or reasoning from evidence, but in the ways these practices and others become intelligible together.

[6]

Narrative Reconstruction, Epistemic Significance, and the Temporality of Scientific Practices

Most discussions of the epistemic significance of narrative have focused on the temporality of the objects of knowledge in specific fields of inquiry. The paradigm cases of narrative knowledge have thus been in history and the human sciences, whose objects of study are human actions, practices, and institutions. If narrative has a place in the natural sciences, it is presumably only in disciplines such as geology, evolutionary biology, and cosmology, whose objects of study are spatiotemporally restricted individuals (planets, species, and so on) that could by analogy be said to be "historical." In this chapter, I argue instead that narrative should occupy a much more fundamental place within our understanding of the sciences, one that addresses the temporality and temporal understanding of scientific practices and hence one that is not limited to specific object domains.

NARRATIVE ENACTMENT

More recent philosophical discussion of narrative as an epistemic category can be traced back to the initial reactions against logical positivism within the so-called analytic philosophy of history. For philosophers such as Louis Mink, a principal concern was to recover the autonomy of historical understanding from the positivists' attempt to assimilate all explanation into the hypothetico-deductive mode. The question of whether narrative had a genuinely epistemic function was decisively shaped by this encounter with positivism in that

> (1) the question was taken to concern only some disciplines, typically history, biography, literature, and perhaps the human sciences more generally, for which the positivist accounts of natural science were thought to be specifically inadequate;
> (2) narrative was taken to be a specific mode of comprehension or form of writing which is or should be employed in those disciplines, and its use was parallel to the role of theoretical comprehension or hypothetico-deductive explanation in the natural sciences; and
> (3) philosophical discussion was focused on whether the structures of the completed narrative text (determinate beginning, middle and end, narrative point of view, and so on) are essential to historical or literary understanding and whether these structures are already found in actions and events or are artifacts of the narrative retelling of them.

Obviously, the turn to narrative opposed the reigning positivism on some important issues, including the unity of science, the universality of the covering law model of explanation, and the epistemological importance of formal logic. But there were also important similarities. Narrative was supposed to be a form of explanation in history and literature parallel to the positivist account of theories and laws in science. This parallel ran quite deep in the work of Mink and Hayden White. They insisted that narrative form was imposed on events, which occur in sequence but do not themselves exhibit any narrative structure. "Stories," Mink insisted, "are not lived but told."[1] The unformed sequences or chronicles that we live through prior to telling stories about them have a similar place in Mink's or White's accounts of historical understanding to that of theory-independent observations in positivist philosophy of science. In both cases, there was supposed to be a strong separation between the form of comprehension and the unformed content on which that form was imposed, and in both cases, this supposition led to problems about how narratives or theories could ever count as true.

There was also, I believe, a deep similarity in the larger projects of positivist philosophy of science and defenses of the autonomy of

1. Mink 1987: 60.

narrative understanding in history and literature. Both programs aimed to legitimate a form of knowledge as an appropriate deployment of a general human capacity. The positivists recognized only one kind of knowledge about the world, knowledge that resulted from the ability to construct and test complex logical constructions (theories) against the results of careful observation. Mink offered one of the most sophisticated defenses of narrative understanding by claiming that science and history instead reflected distinct modes of comprehension. But Mink's account had this in common with positivism (and with the neo-Kantian tradition more generally): the legitimacy and autonomy of disciplines of inquiry are secured as a whole by their fulfillment of the necessary conditions for the exercise of a human capacity.

History and literature no longer seem in dire need of a theory of narrative understanding to fend off the encroachments of positivist methodological imperialism. Yet these early postpositivist discussions of narrative retain a strong hold on the philosophical imagination. The point of drawing out these connections is to clarify just where my concerns and my account of narrative diverge from this tradition and thereby to avoid crucial misunderstandings. At least four crucial differences exist between my account of the narrative dimension of scientific practice and the conception of narrative which emerged from postpositivist philosophies of history and literature.

The first and most obvious difference is that I do not confine the epistemic importance of narrative to a special group of disciplines or objects of knowledge. Narrative understanding is characteristic of any scientific practice, not of particular objects or domains of inquiry. Second, I am not concerned with the form in which the results of investigation are written. I do not want to claim, for example, that scientific papers are really narratives in disguise. Rather, I am interested in the ways in which both the practices of scientific research and the knowledges that result from them acquire their intelligibility and significance from being situated within narratives. Third, narrative should not be thought of as a scheme imposed on an unnarrativized sequence of happenings. Instead, I want to claim that the intelligibility of action and of the things we encounter or use in acting depends on their already belonging to a field of possible narratives. On my view, we live within various ongoing stories as a condition for our being able to tell them or for doing anything else that can count as acting or participating in a practice.

The fourth difference marks a shift away from conceiving the epistemic significance of narrative in terms of completed narratives, with their established beginning, middle, and end and with their unitary point of view. Scientific understanding should instead be situated within narratives in construction (or perhaps better, in continual reconstruction). In the narrative fields of scientific practice, I will argue, from no unitary authorial point of view can an entire course of events be surveyed, for multiple authors engage in an ongoing struggle to determine the configuration of the narrative within which they are all situated. Current scientific research does not simply work out what will happen next within a given domain of knowledge; it continually reconfigures its own past. These four points can be summarized in a single claim: the intelligibility, significance, and justification of scientific knowledges stem from their already belonging to continually reconstructed narrative contexts supplied by the ongoing practices of scientific research.

The claim that scientific research is intelligible only in the context of a narrative extends a central thesis of *Knowledge and Power*. I argued there that science should be philosophically reconceptualized as something that scientists do rather than as a body of representations which results from the activity of research but which afterwards stands on its own. Of course, science has always been recognized to involve activity in the sense that scientists perform experiments, do calculations, construct theories, and write papers. But it is a much stronger claim to say that the intelligibility, significance, and justification of these activities and their products depends on scientists' practical grasp of a research situation as a field of possible activities. I argue that this practical understanding takes on a narrative form, and that through their research activities, scientists attempt to fashion that narrative in a particular way. Much of what happens in day-to-day science concerns the emergence of a coherent narrative field of action from the multifarious doings of different scientists, whose work aims to push the story line in different directions. Only within a largely shared grasp of the current research situation and the possibilities it opens up can the work of particular research groups intelligibly proceed, yet the divergences between the ways the various groups take up those possibilities constantly threatens that shared understanding. Scientific knowledge results from this ongoing tension between narrative coherence and its threatened unraveling.

The preceding chapter offered a much more extensive discussion of practices. In this chapter, my attention is specifically focused on the temporality of practices and especially the temporal dimension of practitioners' understanding of the practices in which they participate. A pattern of events counts as a practice only if it can be construed as having been brought about in part by agents.[2] To attribute agency to someone is, among other things, to understand the agent as acting for the sake of something, but to act for the sake of something is a complex capability. It requires some understanding of how to do many other things, including how to utilize appropriate means, and how to situate the intended action with respect to some further "for the sake of which." A completely pointless or unintelligible action is a contradiction in terms.

The contexts within which actions are both performed and construed as performances are further complicated by being socially situated. The successful completion of actions and therefore the practical understanding by the agent without which they would not count as actions require the action to be appropriately supported by others and to be adapted to contribute in appropriate ways to the actions and purposes of others. This will not happen unless actions are generally performed in the right way, such that they will be intelligible to those with whose activity they must fit in. As Mark Okrent put it, "My end, the for-the-sake-of of my behavior, is impossible as an end unless I belong to a community and the end is a type of end within that community, an end that in turn performs an instrumental role for others within that community."[3] In turn, it is "necessary that there be a certain standardization of ends and of ways of reaching those ends which we share or, at least, which the artificer takes into account in his work."[4]

These brief remarks can be summarized by saying that action must be teleological, holistic, instrumentally mediated, and socially regulated. But these characteristics of action are inescapably temporal. To participate in practices as an agent is among other considerations to be ahead of oneself, that is, to have some understanding of what it would be to have done the action in question. It is also to have some sense of how to initiate or continue the action now. This is in turn to have a grasp of the situation one is already in, to which the action is an intelligible response. These three aspects are held together in the agent's understanding in the form not of an explicit representation

2. For a more extended discussion of the conception of agency and intentionality which is summarized in the next few paragraphs, see Okrent 1988, forthcoming; and Rouse 1987b: chaps. 3–4.

3. Okrent 1988: 51.

4. Okrent 1988: 46.

but of a practical capability. Our projecting ourselves ahead, by taking over the situation we find ourselves already in, by presently acting, is an understanding both enacted and displayed in the action itself.

This unified temporal grasp has a narrative structure. Agents project themselves ahead into a standpoint that could display the temporal field of the action from beginning to intended end. Such understanding in the future perfect tense places the action in a situation from which agents set out and proceed toward an indefinitely intended resolution. What results is not a story told in retrospect but a story that the narrator is in the midst of. David Carr has nicely summarized the sort of view I advocate: "We are constantly striving, with more or less success, to occupy the story-teller's position with respect to our own actions. . . . To be an agent or subject of experience is to make the constant attempt to surmount time in exactly the way the story-teller does, . . . to dominate the flow of events by gathering them together in the forward-backward grasp of the narrative act."[5] Although obviously we often do tell stories, to others and to ourselves, those tellings belong to our ongoing enactment of stories in everything we do.

Understanding narrative as enacted is crucial, for although the narratives we tell do not ordinarily encounter obstacles (there may be many obstacles to the action within the narrative, but not to our narrating them in that way), narrative enactment is continually contested and fragmented. Other people and things are often recalcitrant characters in our enacted narratives; their roles are essential to the projected development of the story, but they do not always pliantly follow the script. Instead, we find ourselves to be quite different characters in the stories they are trying to enact at the same time, with consequences that our subsequent actions (and subsequent understanding of our prior actions) must take into account. As Carr notes, "Sometimes we must change the story to accommodate the events, sometimes we change the events, by acting, to accommodate the story."[6] These actions therefore belong not to a single narrative but to a contested narrative field (that is, a field of overlapping, partially conflicting, and interacting narrative possibilities). What is ultimately at issue in the ongoing accommodations and resistances that make up such a field is just what story or stories we turn out to be, and already

5. Carr 1986: 61–62.
6. Carr 1986: 61.

to have been, in the midst of. These stories are not *about* our world but *of* it. They are not representations of events but the events themselves as significantly configured by their place within a contested story field.[7]

Narrative enactment always takes place from within the middle of the story. Actions are already situated as responses to what came before, but the significance of both situation and action depends on their direction toward possible ends. Those ends are always projected retrospections, cast in the future perfect tense, and open to revision as the situation develops.[8] But such revisions thereby partially transform the sense of the "original" situation and the actions taken within it. What situation we are in and what it is we are doing are therefore not yet fully determined.

Yet if the situation of agents is never fully determined, it is also not indeterminate. To recognize this point, we must realize how my description above of agents enacting conflicting narratives is still misleading because it is too individualized. The intelligibility of agents' own narratives, even to themselves, requires some common space within which they inhabit a world with others.[9] As agents, we conform our own understanding and the actions that arise from it to the narratives that we take to inform the actions of others. These narrative possibilities in turn result from and also embody our understanding of a past that we share with others as participants in various communities. Membership in those communities is itself constituted in substantial part by sharing that past as a basis for further action and by accountability (to ourselves and others) for the intelligibility of those actions in terms of that past.

Unfortunately, however, the recognition that the intelligibility of actions and practices presupposes a sociotemporal location in the shared past of some community or communities names a problem rather than solving it. Suppose we grant that intelligibility depends on an ongoing enforcement (by ourselves and others) of norms of correct or appropriate performance in order to sustain communities

7. It is thus important to regard the narrative structure of human actions not as private, individual mental representations of events but as publicly available possibilities for understanding and responding to situations that people *share*.

8. For a more detailed discussion of this way of understanding narrative, see Carr 1986.

9. As Taylor (1991) cogently argues, that common space cannot be internalized within a particular subject by incorporating the introjected voice of the other within an interior monologue.

with a shared past and mutually intelligible future possibilities. Still, the common narrative that actions thereby implicitly invoke must exhibit a continuing pull toward incoherence. Sharing a communal past that constitutes a present situation as a field of action, different people nevertheless act on that situation in various ways, which develop the supposedly common narrative in different directions. To some extent this variation reflects the fact that different people often belong to multiple communities and situate their actions in a variety of narrative contexts, ones that overlap only partly with those of their colleagues in any one of those contexts.

But follow out these divergences and you also realize that even the members of a community sharing a history have already understood the earlier elements of that history to be configured in subtly different ways. What I originally spoke of as a common narrative is actually more like a contested field of competing narratives. I have in mind something like Alasdair MacIntyre's claim that "what constitutes a tradition is a conflict of interpretations of that tradition, a conflict which itself has (or is) a history susceptible of rival interpretations."[10] Thus, I want to insist against someone such as White that actions already implicitly belong to a narrative field and are not indefinitely open to different narrative configurations or emplotments. But I also want to disagree with MacIntyre, who sometimes writes as if there is one true narrative that most adequately describes a situation with its history and prospects.

Sharing a situation as a narrative field thus makes possible meaningful differences along with convergence. The need to make differences intelligible and a common project possible compels an ongoing struggle to keep in check the divergence of versions of a community's story, even as the various actions of its members strain at the limits of coherent inclusion with one another. This struggle takes the form of a shared concern to construct, enforce, and conform to a common narrative pattern within which everyone's endeavors make sense together. The possibilities for failure and for the collapse into partial incoherence of a community's sense of how to proceed intelligibly are many. But it is important to recognize that when they succeed, it is only through a continuing partial reconstruction of a shared sense of what the community has been about and where it can and should proceed.

10. MacIntyre 1980: 62.

THE SIGNIFICANCE OF SCIENTIFIC PRACTICES

We can now return to the importance of narrative in understanding scientific practice. I will argue that scientific understanding itself has a narrative structure, which is necessary to comprehend how various projects and achievements become epistemically significant. What I call the narrative reconstruction of science is also crucial for understanding the unity and coherence of scientific knowledge, given its geographic, linguistic, and social dispersion. To understand these claims, we need to consider the problems of significance and coherence within ongoing scientific practice.

Significance can be understood as the most fundamental epistemological issue, certainly more fundamental than truth.[11] First, questions of significance arise at all levels of scientific understanding, not just concerning the codification of research results. They include concerns about which projects are worth taking up, what results must be taken into account in working on those projects, which experiments and calculations need to be done, what equipment and skills should be acquired, and which results are worth trying to publish. To understand how these questions are answered is to grasp in considerable detail how scientific knowledge is constructed and legitimated.

But the notion of "significance" can be unpacked to show its importance for the philosophy of science in a different way. The word 'significance' might easily be considered ambiguous; it is used to denote both what is important and what is intelligible. Consider the latter case first. A traditional issue in the philosophy of science concerns whether and how a statement made in the course of scientific work is meaningful at all, prior to the question of whether it is true. The logical positivists unsuccessfully attempted to address this question in a general way with a verificationist theory of meaning. Yet the question can also be addressed historically. Ian Hacking, for example, has suggested that which statements count as true-or-false, that is, as possible candidates for truth which have determinate truth conditions at all, might depend on their relations both to other statements and to the forms of reasoning which bear on establishing their truth or falsity.[12] I would expand Hacking's suggestion in two ways. First, I

11. In making this claim, I do not dismiss the importance of the truth concept. Indeed, I have argued elsewhere (Rouse 1987b: chap. 5) for the indispensability of a minimalist semantic conception of truth for understanding scientific knowledge. The importance of such a conception of truth is developed further in Chapter 8 below.

12. Hacking 1982.

would extend it analogously to actions as well as statements; comparable historical differences exist in what actions could make sense at all as a response to a situation.[13] Second, I would argue that the truth-or-falsity of a statement and, more generally, the intelligibility of a practice, depend not just on "styles of reasoning" and the connections they establish among statements but also on the other practices that form the context of its utterance or use.

The importance of the apparently dual sense of 'significance' becomes clear when we consider where the borderline is between intelligible and unintelligible practices and statements. I take that border area to encompass those statements and practices that are only barely intelligible, because they may have a discernable "meaning" but do not have a point. It includes, for example, those statements whose meaning we can understand in a minimal sense even though we cannot grasp why anyone would ever want to say *that* (or to say that in a particular context). Actions can be treated similarly: for instance, a scientist might well be able to understand (in a limited way) what another scientist is measuring but fail to grasp why the outcome of that measurement could ever matter to scientific knowledge or practice. If this suggestion is correct, the apparent ambiguity of the notion of scientific significance actually represents a continuum from what is unintelligible or pointless to what is of fundamental importance.

Scientific achievements come to be important only within the context of other ongoing scientific work. Thus, questions about the significance of scientific work are connected to questions about the coherence or unity of that ongoing work. Coherence becomes problematic once one takes seriously the dispersion of scientific work. Science is practiced in scattered locations. Its practitioners come from varied cultural and scientific backgrounds, work in different material settings, and often work toward disparate proximate aims. Yet theirs is nevertheless a shared enterprise: research aims to advance investigations already initiated by others and to enable further inquiry in turn. The crucial question is how such dispersion is sufficiently overcome to enable such mutual dependence and to highlight some projects, practices, and achievements as especially significant in the context of that ongoing collective project.

13. What the extension actually does is to subsume the class of true-or-false statements within the larger class of possible (that is, intelligible) actions; this approach is thus of a piece with my more general approach to knowledge as practical engagement with the world and to propositional knowledge as only part of the more complex and heterogeneous embodiment of knowledge.

There is a widely accepted answer within contemporary philosophy of science to the question of how scientific practice achieves coherence and how various theories, experiments, techniques, and so on thereby become significant for the ongoing practice of research. The standard view is that scientific "communities" share fundamental presuppositions (vocabulary, theoretical models, values and norms, and so forth) which are inculcated in their members' professional training and which are enforced by the standard gatekeeping procedures of a discipline. Common commitment to these basic values and theoretical commitments is what supposedly enables research to progress coherently and some projects and achievements to become more significant than others. Various accounts are then provided of how the fundamental commitments of the community change over time (for example, by gradual, reticulated development or by relatively sudden and dramatic shifts in community practice).[14] From the standpoint of the epistemological tradition, this approach has the additional advantage of minimizing and externalizing the social dimensions of scientific knowledge: what matters epistemically are individual cognitive states, and the particular social practices through which scientists come to share those states can be safely left to sociologists.

This standard approach to the issues of coherence and significance is not credible, however, for several reasons. First, I think it has not been borne out empirically. Its proponents have either overstated the degree of consensus within particular scientific fields or have identified shared beliefs and values that are too vague and general to account for the significance relations and coherence within everyday research practice. Second, we ought to be suspicious if a high degree of consensus were identified within scientific communities, because the gatekeeping practices for policing that consensus are not nearly adequate to enforce it.[15] There is considerable slack within scientific training and in the practices of professional review and reward by the sciences, so that many divergent local practices and interpretations are unlikely to be detected and rooted out. Third, the notion of a "scientific community" cannot bear the weight that such community presupposition accounts require. On any plausible account

14. It is worth noting that Kuhn 1970, which is usually taken as the locus classicus on the role of community presuppositions in science, may not belong within the tradition it spawned, because what Kuhn takes to be shared by scientists are not beliefs and desires but practices. For a discussion of this point, see Rouse 1987b: chap. 2.

15. Fuller (1989) has emphasized the absence of systematic checking for whether scientific communities exhibit consensus or a failure to notice or object to dissensus.

of the microcommunities in which everyday research practice is situated, many scientists belong to multiple communities, without the discernable consequences one would expect from simultaneously following subtly incompatible presuppositions. Furthermore, the transactions that do regulate scientists' work and its interpretation are not readily confinable to such autonomous communities of like-minded researchers.[16]

As an alternative to the postulation of presuppositions shared by scientific communities, I suggest that we take the coherence and significance of scientific work to be established by the ongoing reconstruction of contested narrative fields. The crucial question, of course, is what is to be gained by understanding scientific work in this way. In the remainder of the chapter, I will suggest some answers to that question and in so doing will say a little more about what such an understanding of scientific practice would look like. My answer has four basic parts. First, such an account enables us to see how scientific work achieves coherence and significance in a way that allows for conflict and meaningful differences. Second, it makes greater sense of the ways in which the standard representations of scientific knowledge (journal articles, reviews, textbooks, popularizations) are continually reconfigured. Third, it does justice to both the social and the material aspects of scientific practice, neither isolating social "factors" as external influences on knowledge nor collapsing all interactions with things to the social construction of accounts or representations of them. And finally, it gives a new and informative sense to the question of the unity of science.

My account of the importance of narrative for understanding the sciences is first and foremost directed at the question of how scientific work becomes significant. As we have seen, this question concerns both how it makes sense at all to engage in that work and why that work is important. The appeal to narrative is an alternative to the standard view that scientific work becomes intelligible and important against the background of a research community's shared beliefs and desires (even on the standard view, of course, it is recognized that an achievement may acquire significance precisely by creating/discovering a new community whose pattern of shared presuppositions is focused by the new work, but it is precisely such changes that the standard view has difficulty making intelligible). The alternative I

16. For useful discussion of the transcommunal negotiation and regulation of scientific work, see Knorr-Cetina 1981: chaps. 4–5, and Latour 1987: chaps. 3–4.

suggest is that scientific practices and achievements are intelligible if they have a place within the enacted narratives that constitute a developing field of knowledge, and they are important to the extent that they develop or transform those narratives.

An obvious prior question is, Why should we think of scientific work as taking place within a narrative context at all? Narratives of the development of scientific fields are, of course, often constructed retrospectively, by historians, by scientists looking back at their individual and collective accomplishments, and as popularizations of scientific work for a historically conscious culture. But it is a different and much stronger claim to say that the significance of scientific work is situated within narratives that are not merely retrospective but are also constitutive of its sense.

We can begin to understand the rationale for this stronger claim by recalling that scientific work is directed toward a collective enterprise. Work that neither responded to others' investigations nor was taken up and used by others in the course of their own research would at best be marginal to scientific knowledge. The significance of scientific work is thus first and foremost its significance for the course of ongoing research.[17] But that significance must be distinct from the work's conformity to prevailing assumptions, standards, and values within any particular research community. Of course, some work may be consistent with such community presuppositions but be insignificant because it does not advance in scope, precision, or technical competence beyond what has already been done or otherwise fails to have many implications for further possible research. But even once one excludes work that is trivial or localized given previous accomplishments, what is scientifically significant obviously cannot be limited to what is consistent with any prior presuppositions. What such a restriction must leave out is the significance of work that transforms a community's prior commitments or changes what counts as the relevant scientific community.

Any account of scientific knowledge which tries to make sense of the significance of scientific work by appealing to a given background

17. Obviously, some scientific work remains technologically important even after it has ceased to be of scientific importance. But this distinction can be blurred in two ways. First, the longer-term scientific significance of much scientific work also concerns its "technological" applications to instruments, procedures, and materials used in further research. Second, the epistemic dynamics of technological development can be treated in a way quite similar to my treatment of scientific research in a narrower sense. Distinctions can be drawn for various purposes between basic and applied research, science and technology, or research and development, but I think the epistemological significance of those distinctions is minimal.

will confront this difficulty. Whether what is taken as background is some core set of beliefs, standards, methods, or values or is the identity of a community whose beliefs, standards, methods, or values may change over time, such views make some scientific developments intelligible at the cost of making changes in the core background unintelligible. This problem was brought to the forefront by post-Kuhnian debates over the rationality of scientific revolutions and is not really resolved by the now common appeal to "reticulation" or "intercalation" among different parts of a community's commitments,[18] since one must still understand why on any occasion one element of the community's commitments takes priority over others.

The various appeals to presuppositions or communities were intended to explicate how the history of work within a scientific field shapes the significance of subsequent work, by identifying what features of that history are or should be decisive. But what is decisive is not some feature(s) of that history but the history itself as an interpretive resource. The history of research within a scientific field is not already fixed and agreed on. As in most practices, historical understanding in the sciences is closely tied to present issues and conflicts. To appropriate the history of the field successfully can be an important step in establishing a possible trajectory of current and future research practice and legitimating it against plausible competitors; conversely, the actual course of ongoing research reconfigures its history.[19] Thus, what is held in common among practitioners of scientific research is a field of interpretive conflict rather than any uncon-

18. These are the terms introduced by, respectively, Laudan (1984) and Galison (1988) to denote the critique of one part of a community's commitments on the basis of others, which are in turn open to criticism on the basis of still further components. But similar models of scientific change have been put forward by many others as well, differing primarily in how the reticulated or intercalated elements are categorized.

19. This account of how the course of subsequent research transforms what counts as historically significant within scientific practice may seem to resurrect the widely reviled Whig conception of the history of science. But the term 'history' in English, already ambiguous between the interpretive practices and products of historians on the one hand and the various objects of their interpretations on the other, is further complicated by recognizing that historical understanding is not limited to professional historians. I am primarily talking about the historical understanding that informs (literally) current research practice within the sciences (the significance of historical understanding for contemporary scientific research has often been greatly underestimated because of the temporal compression of much of what matters historically to ongoing research practice). But before one makes invidious comparisons between scientists' historical consciousness and historians' more adequate understanding, one must take into account the untenability both of the notion of history *wie es eigentlich gewesen* and of historical understanding as a narrative scheme imposed on an unstructured chronicle. What counts as historically significant for historians itself has a comparable history, which is complexly intertwined with the history they study.

tested commitments about beliefs, values, standards, or meanings.[20] Such conflicts need not be explicit; they may, for example, be evident only in unacknowledged tensions within the ways in which a result is appropriated or a term is used or in unnoticed differences in patterns of citation and use in the course of research.

It may be somewhat misleading, though, to identify the background against which scientific work acquires significance with its history, if that is understood to be its past. The conflicts that constitute a living tradition of scientific research are conflicts over what is to be the future course of research in the field, which are simultaneously conflicts over how to interpret its past. For on the account of research practice which I am suggesting by describing scientific work as a kind of narrative enactment, one configures one's historical situation precisely by drawing on it as a resource for making sense of one's present activities, in a future perfect projection of a course of events within which those activities would count as significant. To engage in one research project rather than another (or more mundanely, to work out some part of that project in one way rather than another), is to (attempt to) reconfigure the story that would make sense of that project within its historical situation.

One might worry that if there were no common content (for example, beliefs, meanings, values, standards, methods, and so on) to account for the unity of a scientific field and for the significance of the various practices and results within that field, these matters would be wholly indeterminate. But there can be a more or less definite configuration to a historical situation, even if there are no uncontested interpretations of what it is. What holds such a situation together is the need to form (and/or conform to) a common situation, that is, for one's own activities to mesh with what others are doing, for them to be epistemically significant. Scientific research always works within a continuing tension between intelligibility and incoherence (in this respect, however, it is rather like any other sort of human activity).[21] The activities of various scientific researchers, left unchecked, would be more or less centrifugal, as each responded to their own

20. Recall Alasdair MacIntyre's elegant rendition that I quoted earlier: "What constitutes a tradition is a conflict of interpretations of that tradition, a conflict which itself has (or is) a history susceptible of rival interpretations."

21. For a somewhat more detailed discussion of how my account of scientific research and knowledge is situated within a more general conception of actions, practices, and meaning, see Rouse 1987b: 58–80. The most extensively developed account of a similar view is to be found in Okrent 1988: Pt. 1.

interpretation of the present state of knowledge. Even small initial differences over what constitutes interesting questions, relevant considerations, important objections, sound technique, and the like would be magnified by the consequences of following out their implications for further work.

The need to align one's work with that of others compels an ongoing struggle to keep such divergence in check, however. Obviously, disciplinary gatekeeping procedures play an important role here, but even more important are the ways in which researchers preempt the need for discipline imposed from without by adjusting their work to (what they take to be) the prevailing epistemic situation. But such continual adjustment should not be mistaken for a return of consensus, for one important form of such adjustment is to challenge important aspects of one's situation (for example, by arguing against them; one important clue to the present situation of knowledge within a scientific field is against which positions and arguments it is necessary to argue and where the burden of proof lies).[22] The identity of a scientific field (or of the present state of knowledge within that field) is established not by any determinate shared presuppositions, but only as a field of intelligible differences, one whose boundaries are themselves contestable. On this account, it is certainly possible for a scientific field to lapse into incoherence, so that it is no longer quite clear what is at issue in pursuing its questions. But the alternative to such incoherence is not a stable core of agreed-on content but the ongoing reconstruction of its history to make a place for present work and its evolving prospects.

Such an account of the ongoing narrative reconstruction of science contributes to an understanding of knowledge as dynamic. Up to this point, I have focused on how that account makes sense of the coherence and significance of scientific work without appealing to reified conceptions of scientific communities, consensus, or meaning (for example, conceptual schemes). In doing so, it makes allowance for the ineliminability of some degree of incoherence and pointlessness within scientific practice. But it also gives reason to reject the reification of knowledge itself, as I shall argue in the next two chapters. It is difficult to identify knowledge with some present body of propositions, sentences, or beliefs if its "content" is temporally extended and deferred.

22. Fuller (1988: chap. 4) discusses the significance of bearing or shifting the burden of proof.

One advantage of such a nonreified conception of knowledge is that it promises a better account of the ways in which the scientific literature is reconstructed in the course of scientific work. As I have argued more extensively elsewhere,[23] neither the scientific literature as a whole nor any of its component parts can be understood as an encyclopedic representation of the present state of knowledge in the sense in which knowledge is typically reified by most epistemologies. The journal literature is obviously not an attempt to represent the present state of scientific knowledge, for it includes contested and conflicting claims. But the various places in which knowledge might plausibly be said to have been collected and reviewed (for example, review articles, handbooks, textbooks, popularizations, or encyclopedias) will not do either. Such collections are always put together for the sake of some subsequent use, which governs the selection and interpretation of its contents. Well-established and subsequently unchallenged results are regularly excised from such collections because they are no longer important for the ongoing practices toward which the collections are directed. Other results are reinterpreted to accord with their prospective use for other projects, a use that may conflict with the original context within which they were established and justified. In many cases, these reinterpretations make previous achievements less precise or nuanced in order to make them more robustly usable. This phenomenon is strikingly evident in the continual replacement of scientific textbooks.[24] Such textbooks become outdated not because their previous contents have been significantly falsified or simply because too many new results have since been achieved (why not just publish occasional supplements?) but because of the ongoing reconfiguration of the significance of even well-established results within the changing state of the field. What the books are about and hence their contents and organization are at issue in ongoing scientific practice.

A much more philosophically consequential reason for serious consideration of this account of the narrative dynamics of scientific knowledge concerns the epistemological significance of the social and material aspects of scientific work and their interconnection. Most philosophical interpretations of science still treat the social situat-

23. Rouse 1987b: chap. 4.

24. The publication of textbooks is actually a more complicated issue, which brings in economic considerations of marketing, production costs, and the benefits of planned obsolescence, the rhetorical complexities of satisfying multiple constituencies, the politics of state certification, and much more. On my account, however, such wide-ranging issues are in no way "external" to understanding the production and justification of knowledge.

edness of knowledge as extrinsic and instrumental to its development at best and of philosophical interest primarily as a possible source of distortion. By contrast, many attempts to attribute a more epistemically constitutive role to social practices and institutions have not attended adequately to scientific interaction with the material world, permitting reference only to socially negotiated accounts of things and never to things themselves.[25] The social constructivist literature has the comparative advantage in this dispute because its criticisms of philosophical and sociological attempts to treat the social aspects of knowledge production as extrinsic to the "content" of scientific knowledge have been largely on target; I will not rehearse the relevant arguments here.[26] But despite more recent discussions of the reflexive application of constructivist arguments, too much of the constructivist literature still both takes for granted its categorizations of the "social" and, worse, premises its arguments on an opposition between what is socially constructed and what is "natural" or material, thus reducing or otherwise subordinating the latter to the former.[27]

The difficulty can be seen in how one might attempt to understand the role of a recalcitrant phenomenon or fact in hindering the development of a research program. Internalist accounts typically focus on the relevant data and the reasoning that would establish their relevance and significance for the program in question, all of which are examined in abstraction from the social and institutional context within which both data and arguments are produced and assessed. Social constructivists aptly respond that data do not speak for themselves and that they can count as evidence against a hypothesis or a research program only if some scientists are prepared to take the data as counterevidence and if these scientists are taken sufficiently seriously by others. Whether these latter conditions hold is then taken to be explainable only by social variables or perhaps a social history.

The inadequacy of this last part of the social constructivist response can be seen from any of several vantage points: the reflexive epistemological arguments frequently raised within the sociological literature itself (How are these "social" explanations themselves

25. For a clear programmatic defense of this social constructivist commitment, see Collins and Yearley 1992a, esp. 314–17.

26. Among the classic studies that display the untenability of treating social aspects of knowledge as "external" are Latour and Woolgar 1986, Knorr-Cetina 1981, Collins 1992, Pickering 1984, and Shapin and Schaffer 1985.

27. It is worth repeating that this objection to "classical SSK" has also been developed forcefully within the sociological tradition itself; see esp. Latour 1987; and Pickering 1989b, 1995.

constructed and validated?), social theories that take seriously the material embodiment of social interactions, or reflection on how the relevant "social" categories can be effectively transformed by new manifestations of worldly things.[28] Insistence on the ongoing narrative reconstruction of scientific practice avoids these difficulties by claiming that the ways social groups form and interact, and the ways things manifest themselves within the context of those interactions, are part of the same story or stories, and neither can be reduced to the other. As Andrew Pickering and I have argued independently, the relevant "resistances" to the achievement and maintenance of epistemic alignments within scientific practice cannot be confined to either social or material categories in opposition to the other.[29]

The final advantage to understanding scientific knowledge as narratively reconstructed epistemic alignments concerns how to conceive of the unity of scientific knowledge across disciplines and research programs. Philosophers have often been concerned to understand the relations among the achievements of different scientific fields. The standard answers are variations on the idea that the telos of scientific work is the unification of its justified representations into a single, consistent whole. This picture of the representational unity of scientific knowledge has both its reductionist versions, in which ideally all knowledge could be captured in a single vocabulary, commonly that of theoretical physics, and its pluralist versions, in which the same things can be accounted for consistently at several irreducible levels of description.

My objection to either version is that it is mistaken in positing the representational unification of scientific knowledge as a goal of scientific practice at all, for reasons put forward earlier in the chapter. Instead, I suggest that multiple, ongoing, narrative unifications of scientific knowledge are part of everyday scientific research practice. Work done in other scientific disciplines regularly becomes a resource for scientific practice (including for the criticism of alternative ways of configuring the prospects for further research). New fields of research are opened by drawing on the concerns and achievements of what once were separate fields. Such uses of other scientific achievements are always partial without being completely decontextualized: when scientists appropriate other work for their own purposes, they often find themselves committed to more than they initially realized

28. This last strategy has been especially prominent in Latour 1984, 1987, 1993.
29. See Pickering 1989b and Rouse 1993.

without being committed either to the entirety of its connected practices and results within its original research context or to the ways they were originally configured. Scientists who appropriate other results often reinterpret or reorganize them, and the extent to which their borrowings must be responsible to their original context may itself be at issue within the new field of work to which they are adapted. Such issues are settled by whether and how the borrowed results can be integrated intelligibly into ongoing research practice in response to the ways in which its integration is contested by those trying to develop the practice in different directions. What results is not a systematic unification of the achievements of different scientific disciplines but a complex and partial overlap and interaction among the ways those disciplines develop over time.

I conclude with some brief remarks about the implications of the overall position I have sketched here for the disciplinary development of philosophy of science. My approach has important affinities and differences with the tradition of naturalizing epistemology and philosophy of science. The parallels include taking the actual development of scientific work as the basis for understanding what knowledge is and ought to be, emphasizing the continuity of such philosophical reflection with empirical studies of how scientific knowledge develops and methodological reflection within scientific practice, and understanding knowledge itself as part of the world we need to understand. But the most prominent conceptions of naturalized epistemology give preeminence to a single scientific discipline (commonly biology, psychology, or sociology) as the successor subject to classical epistemologies.[30] Reflection on the heterogeneity of scientific practices and the multiple and integrated ways in which they are formed, extended, and contested easily suggests the inadequacy of such circumscribed approaches to scientific inquiry as a biological, psychological, or social process.

A more adequate model for a philosophy of science might well be interdisciplinary cultural studies. *Cultural studies* in the sense in which I use the term are directed toward understanding the historical formation and maintenance of meaning. Cultural studies focus on the ways meaning emerges from agents' interaction with one another and their surroundings. This focus does not exclude the ways in which the world materially resists or reinforces efforts to make sense of it:

30. Stump 1992 cogently criticizes the disciplinary monism of most naturalized conceptions of scientific methodology.

the nature/culture distinction is not constitutive of the intended sense of "culture" but is instead a crucial part of what cultural studies need to understand. The direction of cultural studies toward mainly public performance and discourse does not thereby exclude any consideration of cognitive or other psychological processes. The principal theoretical tools of what has heretofore gone under the heading of cultural studies have been political economy, social theory, gender studies, literary interpretation, and interpretive anthropology, but emphasizing the materiality of culture ought to leave ample room for the integration of human ecology broadly construed.

A diverse and growing body of work in science studies, especially by historians, exemplifies what I call cultural studies; I think especially of the work of Donna Haraway, Sharon Traweek, Mario Biagioli, Robert Marc Friedman, Vassiliki Betty Smocovitis, or Paula Treichler. A much wider range of work is also readily assimilable to a cultural studies model. Cultural studies have typically been far more explicitly engaged politically than has been usual within philosophical studies of scientific knowledge. But philosophy of science and science studies more generally has often had a more wide-ranging normative dimension; indeed, many earlier philosophical reflections on scientific knowledge were allied with broader programs of cultural and political reform (think of the manifestoes of the Vienna Circle, or the connections between Popper's philosophy of science and his political philosophy).[31] Otto Neurath especially might be regarded as something of a forerunner to conceiving philosophy of science as cultural studies of the production and utilization of knowledge. In any case, such a model for the interpretation of knowledge seems more promising as a way to do justice (critically as well as descriptively) to the cultural and political complexity of modern practices of knowing than do any of the readily available philosophical or sociological alternatives.

31. Work on the Vienna Circle has increasingly emphasized the connections between its specific philosophical interpretations of science and its broader political concerns; see, for example, Galison 1990; and Cat, Cartwright, and Chang forthcoming. Popper is well known as a political philosopher; the connection between his political philosophy and his philosophy of science is most explicit in Popper 1962.

[7]

The Dynamics of Scientific Knowing: Understanding Science Without Reifying Knowledge

The preceding chapters on scientific practices inevitably pose the question of their implications for understanding scientific knowledge. It might initially seem attractive to regard scientific practices as part of the context or background within which particular claims to knowledge acquire significance and justification. Philosophical interpreters of scientific knowledge now almost invariably recognize that knowledge claims cannot be analyzed singly but instead acquire evidential support and explanatory power only in conjunction with other statements that function as their assumed background. If practices are important for understanding scientific knowledge, then it would seem natural to assimilate them within the postpositivist philosophical tradition by expanding the notion of background assumptions to include practices as well as statements.

Philosophically foregrounding scientific knowledge by relegating scientific practices to the background would nevertheless be a mistake. Or so I argue in this chapter and the following one. The mistake is to reify scientific knowledge and thereby to misunderstand how scientific understanding is dynamic. The recent philosophical literature provides two important models that can guide us in understanding what a dynamic account of science would look like and why it is important. Donald Davidson has argued against reifying languages (or linguistic conventions) as the background to linguistic understanding and communication.[1] Thomas Wartenburg has also argued that social and political power relations are best understood if we avoid reifying either power itself or the social context in which power relations are situated.[2] Neither model should be taken as merely analogous to criticism of the reification of scientific knowledge, for linguistic understanding and sociopolitical power are both integral to scientific practices. We should instead recognize that dynamic accounts of language, power, and knowledge function together in understanding scientific practices and their significance.

My discussion of the import of a broadly Davidsonian philosophy of language for understanding the sciences must be held until the next chapter. In this chapter I use Wartenburg's discussion of power to introduce my conception of scientific knowing as dynamic. I then address some of the philosophical consequences of this conception. In particular, I argue that a dynamic account of scientific knowing is deeply incompatible with epistemological skepticism, relativism, and eliminativism. A better way to think about the epistemological consequences of refusing to reify knowledge is to regard this approach as deflationary. A deflationary understanding of a concept treats it as lacking sufficient theoretical integrity or pretheoretical unity to support substantive generalizations about its instances. A deflationary account thus suggests that 'knowledge' is a useful and learnable term but that it demarcates only a nominal kind. In *Knowledge and Power* I argued for a deflationary treatment of truth; there are many truths, but no nature of truth. Likewise, I now claim, there is much scientific knowledge but no nature of scientific knowledge. Such a deflationary conception affords ample opportunity to interpret or assess particular scientific claims and their justification but no grounds for the legitimation project, that is, for global interpretations and rationalizations or critiques of the scientific enterprise as such.

EPISTEMIC ALIGNMENTS AND THE DYNAMICS OF SCIENTIFIC KNOWING

A dynamic understanding of knowledge may seem very strange at first. Although room for dispute exists about what kind of thing knowledge is, it might seem less readily disputable that knowledge *is*

1. Davidson 1986.
2. Wartenburg 1990.

something possessed by knowers and transmitted or exchanged through communicative interaction. Indeed, the content of knowledge (both what is known and the evidence and reasoning that warrant it as known) may seem to be independent of its particular realizations in texts, utterances, or thoughts and of the specific history through which it came to be known. Modeling the dynamics of scientific knowing on Wartenburg's account of social power as dynamic and situated may therefore be useful initially just in loosening the hold that reified conceptions of knowledge have on the philosophical imagination, thus enabling us to see what it could mean to understand knowledge as dynamic.

Why should we think of social power as "dynamic" in the first place? Wartenburg's account is preceded by extensive criticism of theories that locate power entirely within the actions and understandings of dominant and subordinate agents in isolation from any larger social context. He also insists, however, that his discussion of power as dynamic and situated goes much further than just situating power relations within a social context or background. We might, for example, recognize that whether an agent's attempt to exercise power over another agent is effective may depend on its social context. Following Hume, Wartenburg notes that "the presence of the magistrate functions as an equalizer keeping the armed agent from being able to harm the unarmed agent with impunity."[3] A structural parallel exists between this kind of appeal to a social context in determining whether to attribute to one agent power over another and the familiar appeal to background knowledge in determining attributions of knowledge. In the first case, the interactions between the two agents are sufficient to determine what each agent is trying to do, but the social context determines whether one's actions are actually and relevantly effective in exercising power over the other. In the epistemic case, the appeal to background knowledge would determine whether a knowledge claim is truly and justifiably informative about its intended objects.

Wartenburg finds two fundamental reasons why acknowledging the "contextuality of power attributions" does not sufficiently capture how power relations are situated in a larger relational field. First, the situation in which agents act determines not merely whether one effectively exercises power over another but also where and how the power effects of those actions are directed, that is, on whom they act and in what ways. Thus, Wartenburg insists, "many relationships of

3. Wartenburg 1990: 148; the reference is to Hume 1888: 312.

social power are constituted *in the first instance* by the way in which peripheral social agents treat both the dominant and the subordinate agents."[4] Second, the context or background against which power relations are thus constituted is not something already determined but is itself continually reconstituted by agents' ongoing activities and responses to one another, including those being foregrounded against that "context." Thus, whether there are power relations among various agents' actions depends on how those actions are situated, but the situation is itself reconfigured by those actions.

Wartenburg tries to capture this dynamic situatedness of power relations in terms of their mediation by "social alignments." This concept is crucial to my adaptation of Wartenburg's approach, for I will shortly introduce the parallel notion of an "epistemic alignment" to understand the dynamics of attributing scientific knowledge and contesting such attributions. Wartenburg first characterizes social alignments as follows:

> A field of social agents can constitute an alignment in regard to a social agent if and only if, first of all, their actions in regard to that agent are coordinated in a specific manner. To be an alignment, however, the coordinated practices of these social agents need to be comprehensive enough that the social agent facing the alignment encounters that alignment as having control over certain things that she might either need or desire. . . . The concept of a social alignment thus provides a way of understanding the "field" that constitutes a situated power relationship *as* a power relationship.[5]

On Wartenburg's view, a relationship between two agents is a power relationship only because and to the extent that other agents will normally respond to the two agents' actions in ways that are coherently aligned with the dominant agent's actions. For example, when judges sentence prisoners or teachers grade students, they exercise power only when other agents (police, bailiffs, and prison staff in the one case; potential employers and school admissions officers in the other) act or are prepared to act in ways oriented by what the judges or teachers do.

The alignment of these actions is not simply a causal consequence of a discrete action by the dominant agent, but instead makes up an

4. Wartenburg 1990: 149.
5. Wartenburg 1990: 150.

ongoing relationship within which each agent's actions are situated. The actions of the judge or the teacher would not make sense without the prior presumption that others would align their subsequent actions accordingly. Indeed, the ways in which other actions are aligned with those of the judge or the teacher help determine what action the judge or teacher actually performed. The judge may intend to help rehabilitate the prisoner and the teacher to educate the student, but others' actions may be coherently aligned with theirs in ways that thwart or transform their intentions (these effects in turn may or may not coincide with the intentions of the aligned agents; power always involves intentionality, but it need not be explicitly intended). Thus, the alignments that constitute situated power relationships need not be governed by the intentions of any of the agents who belong to it or by any other prior characteristic of their actions. Agents can even exercise power directly contrary to their intentions, so long as other agents orient their actions in response to them.

The claim that power operates only through such social alignments has several important consequences. First, power is recognized to be dispersed and heterogeneous. Power is not simply possessed and exercised by a dominant agent or even located specifically within her relationship to a subordinate agent. The exercise of power is mediated by the actions of many peripheral agents, ones that often establish or enforce the connection between the dominant agent's actions and the fulfillment or frustration of the subordinate agent's desires. Social alignments thereby extend quite far beyond the immediate context in which the exercise of power is focused, for they include not only the various ways in which others' actions are specifically oriented by what a dominant agent does but also the availability or absence of alternative access to the goods blocked or enabled by the exercise of power. Large-scale social structures may thus be implicated in what is apparently a local and specific exercise of power.

A second important consequence is that power is dynamic and spread out over time. As Wartenburg notes: "The *present* actions of a dominant agent count on the *future* actions of the aligned agents being similar to their *past* actions. But this faith in a future whose path can be charted entails that the dominant agent not act in a way that challenges the allegiance of his aligned agents, for only through their actions can that future be made actual."[6] This consideration explains why the exercise of power does not consist simply in one

6. Wartenburg 1990: 170.

action forcibly constraining or modifying another. It is partially constituted as a power relation by its reproduction over time as a sustained power relationship. But in this way, social alignments also constrain as well as constitute the exercise of power, and they offer opportunities for subordinate agents to resist or evade power by disconnecting it from its supporting alignment or by constructing countéralignments.

Wartenburg thus presents the social world as a complex array of overlapping social alignments whose orientations constitute ongoing struggles of domination and resistance within which individual agents and their actions are situated. Power is therefore not a thing or commodity that agents possess or exercise; it is dynamic and hence never simply present in specific actions and their effects. It is also heterogeneously embodied in people, institutions, practices, social structures, and (I shall argue) the things people act with, on, and among. Power exists only through being reproduced, and it is continually reshaped by the ways its reproduction is resisted.

We can begin to see how to model scientific knowledge on Wartenburg's dynamic, situated conception of power by considering the importance of experimental practices in making aspects of the world accessible to scientific interpretation. Historians, philosophers, and sociologists of science have in the past decade devoted considerable attention to experimentation and fieldwork.[7] It is now widely recognized as significant that the proximate objects of scientific interpretation are the effect of extensive preparation and manipulation, often requiring complex instrumentation, delicate skills, and extensive practical familiarity with materials, instruments, and procedures. The creation of phenomena[8] or of inscriptions that serve to indicate phenomena[9] rearranges the world in ways that make it amenable to scientific understanding.

At least initially, the creation of a phenomenon is a highly localized activity. Scientific practitioners develop specific "experimental sys-

7. Among the most prominent works in the turn to experiment are Hacking 1983, Ackermann 1985, Pickering 1984, Gooding et al. 1989, Galison 1987, Shapin and Schaffer 1985, Knorr-Cetina 1981, Latour and Woolgar 1986, Traweek 1988, Franklin 1987, and Rouse 1987b.

8. Hacking 1983: chap. 13; the elements involved in creating phenomena are more extensively analyzed in Hacking 1992.

9. Latour and Woolgar 1986 and Latour 1987 have emphasized the importance of inscriptions as constitutively indicative of phenomena and objects; one might say that the difference between Hacking's 1983 discussion of phenomena (publicly manifest regularities in the world) and Latour and Woolgar's discussions of textual inscriptions (marked, if you like, by a high signal/noise ratio) is only the result of the latter pair's predilection for "semiotic ascent." Hacking (1988) provides his own discussion of this parallel.

tems,"[10] using locally situated materials, protocols, techniques, and skills to produce phenomena (or their manifestations) that do not exist anywhere else. To reproduce them (or their analogues) in other places always requires some local adaptations and may, at least initially, require extensive simulation of the original context or even the physical transfer of personnel and materials. Nevertheless, these highly localized activities usually make sense primarily as practices that aim to be informative about objects and events that do or might occur elsewhere. The gradual adaptations, reproductions, extensions, and standardizations of these localized activities needed to realize this aim constitute "epistemic alignments" in a sense analogous to Wartenburg's social alignments that mediate power: they enable these localized activities and achievements to be informative just as power alignments enable particular constraining or coercive actions to be effective. Moreover, they do not merely determine whether they are informative but also what they are informative about, in what ways, and with what significance. Some classic studies of particular cases illustrate this point well: Ludwik Fleck's study of the Wassermann reaction can be understood as working out how locally variant protocols for standardly identifying a condition of blood came to be informative about the historically interpreted disease entity syphilis, and an expansion of Bruno Latour and Steve Woolgar's discussion in *Laboratory Life* could comparably characterize how manipulations of a synthetic peptide became informative about the brain processes that regulate release of the hormone thyrotropin from the thyroid.[11]

Some elements of such confusing arrays of localized projects, practices, and capabilities turn out to reinforce and strengthen one another and are taken up, extended, and reproduced in various new contexts. Others remain isolated from or in conflict with these emergent strategies and gradually become forgotten or isolated curiosities. Thus various patterns and directions to scientific activity gradually emerge, and these provide a setting in which further inquiry can be intelligibly organized.[12] I take these strategic alignments to be consti-

10. Rheinberger 1994, and Kohler 1994.

11. Fleck 1979; Latour and Woolgar 1986. For their own purposes, Latour and Woolgar themselves tell a more circumscribed story about the identification of thyrotropin-releasing hormone, but Hacking (1988) indicates some of the other elements whose alignment was crucial to the ongoing story of how synthesis and examination of Pyro-Glu-His-Pro-NH_2 came to be informative about brain processes.

12. Hans-Jörg Rheinberger's (1992a,b,c, 1994) studies of the tangled emergence of experimental systems and disciplinary research programs in cell biology provide an especially clear illustration of these claims. The much studied case of the multistranded development of molecular biology as a discipline is also a revealing case. See esp. Abir-Am 1985 and Kay 1993 for a sense of the nonunilinear elements of this history.

tutive of what can count as knowing. Thus, a statement, a skill, or a model does not acquire epistemic significance in isolation or instantaneously, depending instead on its relations to many other practices and capabilities and especially on the ways these relations are reproduced, transformed, and extended.

On this conception, the achievements of the isolated, solitary knower are not merely constrained by the limits of any one person's cognitive capacities but are more fundamentally marginal to knowledge precisely because of his or her isolation. A belief, skill, model, or practice gains its epistemic significance from being taken up, refined, and extended into new contexts. Thus, Mendel's work on hereditary transmission in pea plants had a significant place within mid-nineteenth-century hybridization studies, became marginalized as this original context became isolated from other ongoing research projects, and acquired a new and heightened significance through its early-twentieth-century "rediscovery."[13] Mendel's work was further redefined by the development of population genetics in the 1920s and its incorporation into the neo-Darwinian evolutionary synthesis. The differing ways in which the "same" work was taken up and used obviously affected its epistemic significance but also to some extent transformed what the work itself had achieved (that is, what its content was). The content of knowledge, what the work "says," is always what it says or could say to specifically situated audiences.

Yet one should not take this point as emphasizing the social aspects of knowledge as opposed to its referential character. The alignments that constitute knowledge include relations to coworkers who might use, extend, challenge, or ignore one's work, but also to the objects and materials encountered within that research. George Beadle and Edward Tatum's early work in biochemical genetics depended on the significance of questions about the mechanisms of gene action for those biologists already situated in the tradition of chromosomal genetics *and* on their ability to establish an experimental system *(Neurospora)* in which mutations could be rapidly introduced and their metabolic products detected biochemically within a simple culture medium.[14] By contrast, the possibilities opened up by the *Neurospora* studies helped marginalize the work on *Paramecium* which Sonneborn and his students had been developing with considerable success for

13. For a discussion of the significance of Mendel's work in its original context and how its place within biological science was subsequently transformed, see Olby 1979, 1985.

14. For a useful summary of Beadle and Tatum's work, see Allen 1978: 198–205.

several decades, because they were unable to grow the organism on a uniform medium and thus unable to connect their results to the new questions about biochemical mechanisms.[15]

These examples should recall the themes of the preceding chapter by suggesting how epistemic alignments have a complex temporality. A research project typically is addressed to the present situation in a scientific field, including both its past achievements and the prospects and opportunities for further development. Its significance and standing depend on other research continuing to develop in ways that support and build on it (or alternatively, on the course of research changing, whether anticipated or not, in ways that constructively engage it). Thus, some scientific work can become marginal to or even excluded from scientific knowledge, not because it is false, but because it does not fit in appropriately with ongoing research (needless to say, such exclusions and marginalizations are to some extent reversible, as the well-known examples of Mendel and Barbara McClintock attest). Knowledge (in the form of epistemic alignments) is best understood not as a system of propositions or a cognitive state but as a situation in the world. A situation is not fully determinate in the present (it is not just that we do not yet know what situation we are in but also that it is not yet determined what our situation is). In this sense, whether a body of scientific work is knowledge is metaphysically analogous to whether an event is "the decisive turning point" or whether a runner crossing the plate in the second inning is "the winning run."[16]

We need to be careful, however, in claiming that practices do not constitute knowledge by themselves but do so only through the ways that they coordinate with other practices or with specific elements or features of other practices. The reason for caution is that the concepts of 'practice' and 'alignment' are not altogether distinct. Practices themselves are composed of patterns of ongoing activity, so that there exists a practice only if some actions "align" themselves to constitute such a pattern. Furthermore practices are not sharply individuated and in any case can combine with one another to form larger or more complex practices. Various practices of fixing and staining biological specimens, for example, themselves belong to the practice of using light microscopes to produce magnified images, which in turn belongs

15. For extensive discussion of Sonneborn's work and its transformed significance in the light of biochemical genetics, see Sapp 1987.

16. I owe this analogy to Wheeler 1991.

to a tradition of experimental practice within the biological sciences. The reason for introducing the concept of an alignment is that this notion of a coherent, ongoing 'practice' still connotes more temporal continuity and less heterogeneity than is often displayed in the ways practices and their elements come together to make up scientific knowledge, or so I shall argue in what follows.

We should be similarly careful about the notion of an *epistemic* alignment. I invoke this term not to suggest that there are aligned elements or ways of aligning that are *distinctively* epistemic but simply to indicate an interest in those patterns and interconnections in the world which enable some things to be informative about others. Just as Wartenburg wants his account to allow for new possible ways in which agents' actions can function together effectively to constrain or transform the "action-environment" of others (or to resist such effects), part of my concern will be to emphasize the open-endedness of the ways in which agents and their surroundings can be configured to reveal new aspects of the world or to challenge what has previously passed as genuinely informative practices and achievements.

The heterogeneity and open-endedness of epistemic alignments is crucial in distinguishing my account of scientific knowledge as located in the dynamics of the ongoing practice of research from the ways philosophers of science have often tried to account for the "holism" of scientific interpretation and justification. Holistic accounts of knowledge have emphasized that some sentences (or beliefs) count as epistemic justification for others only against a presupposed or otherwise accepted body of "background knowledge" relevant to a scientific field or domain. But the typical invocations of a background make it both homogeneous (either a network of widely accepted or presupposed beliefs or of previously justified sentences) and relatively stable.[17] Instead, my claim will be that the "field" within which scientific claims and practices acquire their significance and justification involves many things that cannot be reduced to sentences or beliefs "internal" to a domain of investigation: skills and techniques, instruments and material systems (including networks for their manufacture and supply), availability of resources (money, of course, but also staff, information, an audience, and so on), institutional structures, relevance to other social practices or political con-

17. A sophisticated example of such an attempt to account for the heterogeneous considerations that shape scientific knowledge as a unified field of "background knowledge" can be found in Shapere 1984.

cerns, and much more. The background against which knowledge claims are interpreted, used, and assessed is also not invariant across the contexts in which they show up. This variance is certainly characteristic of the ways the "same" statement (or concept, technique, device, and the like) is employed within different disciplines, but even within the same discipline, considerable local variability exists in what is taken to be epistemically relevant background. Such backgrounds also change over time, especially as the direction of research shifts.

My emphasis on the heterogeneity of the practices that compose scientific knowledge may initially seem to depart from Wartenburg's conception of "social alignments," for Wartenburg restricts the domain of power to social relations among human agents (and their social institutions and practices).[18] As I have indicated, I reject those conceptions of scientific knowledge which similarly locate it exclusively within networks of social interaction. But Wartenburg is mistaken in imposing this restriction on the composition of the alignments constituting power, because power is often mediated by the environment within which it is situated and by physical things within that environment. This point has important implications for understanding the interconnections between knowledge and power.

Power relationships depend not only on maintaining an alignment of other human agents, but increasingly on a reliable physical environment and the things, processes, and interactions it contains. For example, power relations among gay men, insurance companies, employers, and the state have been significantly affected by the ready availability of the Eliza and Western Blot procedures for identifying antibodies to HIV in blood.[19] The availability, reliability, and significance of these laboratory protocols have raised or transformed many politically charged questions. These include the medical diagnosis and/or political self-identification of gay people and of people with AIDS and the cultural politics of representations of sexuality and sexual identity; they also include the anonymity and confidentiality of testing and its relation to discrimination or oppression, epidemiology, pharmaceutical intervention, and the emergence of an "AIDS service industry." The previous questions have in turn changed the significance of legal protection for the civil rights of gay men and lesbians and of those "disabled" by HIV. It would be highly misleading to

18. Wartenburg 1990: 3–8.

19. Patton (1990) and Treichler (1988, 1991, 1992) offer wide-ranging discussions of these issues which address in considerable detail the ways in which testing reconfigures relations of power and knowledge in medical, sexual, and political identifications.

limit the dissemination of power through these relationships to narrowly social alignments.

Other prominent examples of things or processes that mediate power and resistance readily come to mind. The Europeans who invaded and colonized the Americas and Australasia were accompanied by microorganisms that laid waste to native populations and by domesticated animals and plant species that dramatically transformed the ecology to which native ways of life were adapted.[20] The alignments of power that confronted the inhabitants of those regions and dramatically changed their life possibilities could not be understood as just social relations to other human agents. An apparently quite different example, the various technologies of surveillance and confinement, highlights the point that agents and things are often partially interchangeable elements within a power alignment, so that it becomes difficult to justify the analytical separation of their roles in constituting power.[21] Military force, industrial production, and the medicalization of human bodies offer many other salient examples of power relationships that cannot be understood solely in terms of the aligned actions of human beings. Thus the alignments through which power is exercised and extended but also resisted or evaded must include our physical surroundings and the tools, processes, practices, and effects of the things with and on which we work.

It is not hard to see why theorists of power have not wanted to include the behavior of things within the alignments that constitute "social power." Ever since Marx's critique of the fetishism of commodities, social theorists have been alert to the dangers of reifying social relations or of replacing them with relations among things. But by treating the physical world as merely congealed social relations and thereby restricting alignments of power to the actions of specific agents, one overlooks the ways in which things can resist effective inclusion within those alignments. Things break down, are unavailable when needed, convey confusing signals, and sometimes even get in the way. Things can also open new possibilities for resistance to the

20. On the role of microorganisms in the European conquests, see McNeill 1976; on the significance of transplanted animal and plant species in the European expansion, see Crosby 1986. Crosby claims that "if the Europeans had arrived in the New World and Australasia with twentieth-century technology in hand, but no animals, they would not have made as great a change as they did by arriving with horses, cattle, pigs, goats, sheep, asses, chickens, cats, and so forth" (173).

21. Foucault 1977 is the classic study of this case, but newer technologies and applications (for example, electronic monitoring of parolees) further highlight the points Foucault made about the architecture of visibility.

power relations they mediate. And when things do fall out of alignment in these ways, the effects on power relationships are quite comparable to those which follow the breakdown of social alignments. We avoid fetishism not by strictly separating the natural and the social or by reducing the natural to the social but by recognizing the artificiality of the distinction.

My central thesis in *Knowledge and Power* was that no relevant boundary exists between the "natural world" and the "social world" for understanding power. Power does not refer only to direct causal effects of one agent's actions on another's; it includes the effects of action on the configuration of practices and things within which an agent's actions are intelligible.[22] An important aspect of power relations is their mediation by things, and an especially important feature of power in the world today is its mediation by materials, things, processes, and practices that originate or play an important role in the construction and extension of the "microworlds" of scientific laboratories.[23] We do not need to ascribe interests, values, rights, or a preferred "natural" order to material things to recognize their integral engagement in the operation of power. The point is not to claim that everything belongs to alignments of power but to deny that anything is excluded from them in principle.

But here our principal concern is to introduce and explicate the notion of an epistemic alignment. The claim that epistemic alignments are heterogeneous emphasizes the dispersion of knowledge across complex alignments of practices, environments, and the agents and things situated within them. As John Hardwig has pointed out, this characteristic of scientific knowledge has become more apparent with changes in the scale and complexity of scientific knowledge, as a single experimental demonstration or mathematical proof may depend on a range of skills and judgments that no one individual is competent even to assess, let alone command.[24] Parallel changes in research practice require technically complex and expensive instrumentation merely to gain access to the objects of scientific study. Yet such developments are differences in degree, not in kind, or so I shall argue. The point of making this claim is not to deny a role to individual agents or knowers in establishing or sustaining knowledge but

22. On this notion of a "configuration" of practices, see Rouse 1987b: 62–65, 73–78, 182–85.

23. On microworlds, see Rouse 1987b: chap 4, esp. 101–8; on the inseparability of the social and natural, see Rouse 1987b: chap 6–7.

24. Hardwig 1991.

only to show that their role is much more situationally dependent than is often acknowledged. The extent of this dependence becomes clearer when one recognizes, alongside the dispersion of knowledge, its temporal deferral, which I prefer to characterize as the dynamic character of a scientific knowledge.

My suggestion that the "background" to scientific work is continually shifting thus reiterates the point in previous chapters that knowledge has a more complex temporality than do sentential networks or collections of operant or abiding warranted beliefs. In saying that knowledge is temporally diffused or deferred, I claim that for a statement, model, technique, or practice to count as knowledge, it must be taken up as a resource for various kinds of ongoing activity (whether in further research or in various "applications" of knowledge).[25] Knowledge in this sense continually circulates (another way of saying that it exists only through its ongoing reproduction). Even the various points at which it is articulated, or collected and assessed, are caught up in that circulation. What is proposed as possible new knowledge, whether in informal discussion or initial publication, has an element of tentativeness about it. What is gathered together in retrospective judgment is always oriented toward a further advance and shaped by that projection. The scientific literature itself is always continually reorganizing what is known as a resource for further investigation.

On this account, epistemic conflict is especially important to the temporality of scientific knowledge. Once it is recognized that knowledge exists only through its reproduction and circulation, the importance of conflict or resistance becomes evident: resistance focuses and directs that circulation. The extension and articulation of knowledge (and hence its realization as knowledge) typically encounters various kinds of resistance. And it is precisely in those respects that researchers will try to develop and articulate their work most extensively and precisely. Where there is (possible) resistance, new and more powerful techniques will be sought, more precise and careful measurement will be provided, and theoretical models will be refined to eliminate or bypass possible sources of inaccuracy or unrealistic assumption. These various refinements are themselves new knowledges and in turn often provide further new directions or problems for research. Hence, from around the specific points where knowledge is resisted

25. On the conception of knowledge I am trying to articulate here, the word 'application' is actually somewhat misleading, because we do not first gain knowledge, then apply it; something counts as knowledge only through the ways it is interpreted and extended in use.

emerges a whole cluster of new local capabilities and their extension into new contexts. But the contrary is also true: where knowledge goes unchallenged, where a claim "goes without saying," one finds little or no articulation or development. And where previous resistance vanishes, knowledge also ceases to proliferate.

The relevant forms of "resistance" to knowledge cannot be easily reduced to traditionally narrow epistemic categories. Obviously, knowledge can be resisted because there are gaps in the data, dubious assumptions in the theoretical models, countervailing evidence, or conflicts with other work. But it can also be resisted because the procedures and capabilities for its articulation and development are too expensive, environmentally unsound, cruel to animals, politically sensitive, of too little or too much interest to the military, unprofitable, and so forth. It is precisely here that philosophers have often tried to distinguish considerations internal to knowledge from those which impinge on it from the "outside." Yet the recognition of the locality of scientific work and its dynamics undercuts any attempt to make such a distinction. The sense of a claim and the ways in which it is articulated and deployed in further research and development depend on considerations that do not respect internalist boundaries. All the small local decisions about research materials, equipment, procedures, funding, personnel, skill development, and the like shape the actual development of the more various kinds of knowledge which invest and underwrite the sorts of knowledge claims that philosophers typically investigate. The actual justifications offered for these decisions usually interchange and balance supposedly internal and external considerations. Thus, a physicist might argue for a particular experimental strategy against its competitors by claiming that it is cheaper, would likely produce more relevant detectable events, takes best advantage of the skills of available personnel, might interest new funding sources in the military, is more reliably established in the literature, could more readily accommodate real-time analysis and adjustment, would be adaptable to a variety of experiments, is more elegant in design, and would be less vulnerable to the vocal objections of local environmental groups. My point is that such heterogeneous concerns and reasons function together in the actual contingent shaping of knowledge.

This emphasis on the positive role of conflict and resistance in developing scientific knowledge indicates an important advantage provided by conceiving of scientific knowledge in terms of "epistemic alignments." It enables us to understand the ways scientific practices sustain and reinforce one another to constitute knowledge without our having to postulate an underlying community of scientists, a scientific consensus on theoretical or methodological presuppositions, or a multiply realizable "content" of knowledge contingently embodied in those particular practices. Wartenburg's conception of power alignments recognizes that agents may act in concert to exercise power over others within a particular setting without necessarily having shared intentions, beliefs, interests, or values and without thereby constituting a community in any philosophically or politically interesting sense. More subtle, Wartenburg's conception of power does not exclude ongoing resistance to power; power and resistance are dynamically related rather than diametrically opposed such that power is exercised only to the extent that effective resistance to it is overcome. Similarly, the embodiment of knowledge in epistemic alignments recognizes that knowledge is dynamically related to the various kinds of resistance posed by anomalies, inconsistencies, disagreements, and inadequacies of skill, technique, and resources. Knowledge is understood as an aspect of ongoing engagement with the world and not as a project that could in principle be brought to completion. Just as Wartenburg's account disentangles the notion of power from that of unconstrained domination, we can likewise disentangle the notion of knowledge from the twin telos of completeness and certainty that has haunted the epistemological tradition. In this sense, conceiving of knowledge as disseminated through epistemic alignments belongs to the broad pragmatist tradition that identifies knowledge with the ability to cope successfully with the knower's environment and historical situation.

Such recognition of the dynamics and heterogeneity of scientific knowledge returns us forcefully to the issue of epistemic significance raised in the previous chapter. We cannot presume that knowledge about the (natural) world—or the search for such knowledge—is a self-contained and autonomous domain of human concern. What knowledge is depends in significant part on how practices of inquiry and criticism are situated among other practices and what successful engagement in inquiry is supposed to be *for the sake of*. Historical studies by Mario Biagioli and by Steven Shapin and Simon Schaffer highlight this point by suggesting that supposedly familiar and exemplary cases of "modern" science cannot so readily be assimilated to current presumptions about the nature and significance of knowing nature.[26] Biagioli argued that the dynamics of courtly patronage significantly shaped what it could mean to make knowledge claims, what

26. Biagioli 1993, Shapin and Schaffer 1985, and Shapin 1994.

it was for claims to be credible, and what was at stake in assessing competing claims. Galileo's self-fashioning as a courtly philosopher also refashioned what it was to know and how things could manifest themselves as objects of possible knowledge. Shapin and Schaffer tried to reconstruct some debates over the nature and significance of natural philosophy which surrounded the formation of the Royal Society and to show the conceptions of "social order" within which competing conceptions of inquiry and knowledge were intelligible. Both cases strongly suggest the misguidedness of attempting simply to assimilate the work of Galileo, Hobbes, or Boyle to more recent conceptions of what it is to do science or to aim for and achieve knowledge of the natural world.

These disputes and differences over the significance of knowing nature cannot be confined to the early history of modern science, as much work in science studies amply testifies. The Morrill Act, the Flexner Report, the Rockefeller Foundation program to emancipate the study of "vital processes" from servitude to agriculture and medicine, and Julian Huxley's interpretation of the evolutionary synthesis display fundamental differences over the nature and significance of biological knowledge.[27] The vast changes since the Second World War in the material, financial, and social scale of research in particle physics also thwart any attempt to assimilate them as increasingly powerful and complex means to a common or continuous end.[28] Robert Marc Friedman's account of the construction of *a* modern meteorology, Donna Haraway's studies of what has been at stake in twentieth-century knowledge of primates, and Paul Rabinow's speculations concerning the emergence of "biosociality" as the locus of genetic practices in contrast to prevailing conceptions of "society" and social knowledge: all strongly suggest that the nature and significance of knowing are very much at stake in the ongoing reshaping of scientific practice.[29]

DEFLATING 'SCIENTIFIC KNOWLEDGE'

The preceding discussions of the dynamics and heterogeneity of scientific knowing pose interesting and far-reaching questions for

27. On the Rockefeller Foundation, see esp. Kohler 1991 and Kay 1993; on Huxley, see Smocovitis 1992, forthcoming.

28. For discussions of the significance of these changes, see esp. Galison and Hevly 1992; Traweek 1988, 1992; Galison 1985, 1987, forthcoming; and Pickering 1989a.

29. Friedman 1989, Haraway 1989, Rabinow 1992.

philosophical discussions of the nature of knowledge or of scientific knowledge specifically. What implications does my account of the dynamics of epistemic alignments have for more traditional philosophical conceptions of knowledge, as justified true belief, truth-reliable relations to facts, explanations for the instrumental success of scientific methods, internalization and growth of domain-specific reasoning and background beliefs, and so forth? Michael Williams trenchantly brings these questions to a focus: "We can talk of 'our knowledge of the world,' but do we have any reason to suppose that there is a genuine totality here and not just a loose aggregate of more or less unrelated cases?"[30]

Williams's insistence on the unnaturalness of philosophical skepticism offers a useful way to approach the philosophical implications of understanding scientific knowledge as dynamic. Williams identified a crucial but undefended presumption underlying epistemological skepticism, namely, that the category 'knowledge' demarcates a coherent object of philosophical inquiry. If the project of assessing whether any beliefs genuinely qualify as knowledge is to make sense, then the category 'knowledge' must have sufficient theoretical integrity or pretheoretical unity to permit it to be surveyed and assessed as a whole. As Williams notes, the skeptic (or the epistemologist who takes seriously the need to refute the skeptic) does not forego assessing beliefs one at a time merely for reasons of convenience; only the wholesale assessment of coherent groups of beliefs can carry the force and generality that would make skepticism a philosophical issue. Hence both skeptics and their philosophical opponents typically begin by presuming either the topical or epistemological integration of knowledge (or subcategories such as "knowledge of other minds," "knowledge of the external world," or in the case I am considering, "scientific knowledge").[31]

Deflationary interpretations of 'truth' offer a useful analogy to counter such epistemological presumptions of the unity and integrity of the category 'knowledge'. Deflationary accounts of truth recognize that there is a well-established practice of ascribing or imputing 'truth' but see no need to ground or explain that practice by a theory that would characterize the underlying "nature" of truth. Arguments for deflationary or "minimalist" approaches to truth have been made

30. Williams 1991: 102.

31. A body of knowledge would be topically integrated if all its constituent statements or beliefs made up a coherent, systematic whole; it would be epistemologically integrated if it depended on a common source of justification, for example, "the senses."

in a number of ways. Behind all these arguments is a widespread dissatisfaction with any of the available proposals for a substantive theory of truth: if correspondence, coherence, or pragmatist "theories" of truth had been developed in satisfying and informative detail, deflationary approaches would undoubtedly have been less appealing. Nevertheless, deflationists have motivated their minimalist conceptions of truth in several more constructive ways that may usefully inform our reflections on scientific knowledge.

An important strategy for motivating deflationary accounts of truth has been to emphasize the disanalogy between 'truth' and terms that clearly denote properties. Paul Horwich aptly characterizes the temptation to make such analogies:

> Just as the predicate, 'is magnetic', designates a feature of the world, *magnetism,* whose structure is revealed by quantum physics, and 'is diabetic' describes a group of phenomena, *diabetes,* characterizable in biology, so it seems that 'is true' attributes the complex property, *truth*—an ingredient of reality whose underlying essence will, it is hoped, one day be revealed by philosophical or scientific analysis. The trouble is that this conclusion is unjustified and false.[32]

Horwich rejects this conclusion because he takes the truth predicate to be syncategorematic. Instead of denoting a substantive property of things (for example, of sentences or propositions), it serves a purely semantic function, namely, to permit a speaker to express an attitude toward a proposition (sentence) without having to express the proposition itself. The truth predicate enables one to say that everything Helen said about a particular event is true or that there must be something true in what David said even if it is unclear exactly what. Approaching deflationary views in this way emphasizes the importance of having available the Tarski schema for material equivalence —'*s*' is true if and only if *s*—to capture the semantic function of the truth predicate. Deflationary accounts of 'truth' would thus in the end presuppose the possibility, at least in principle, of doing without the truth predicate by simply asserting the relevant truths wherever they can be identified.

Williams, Arthur Fine, and I have offered a more expansive approach to accounting for why 'truth' should be understood in a deflationary way.[33] What matters on our accounts are the irreducible open-

32. Horwich 1990: 2.
33. Williams 1986, Fine 1986c, Rouse 1987b: chap. 5.

endedness and context sensitivity of appropriate applications of the truth predicate. Consider Fine's apt summary of this deflationary strategy:

> The concept of truth is the fundamental semantic category. Its uses, history, logic, and grammar are sufficiently definite to be partially catalogued, at least for a time. But it cannot be "explained" or given an "account of" without circularity. Nor does it require anything of the sort. The concept of truth is open-ended, growing with the growth of science. Particular questions (Is this true? What reason do we have to believe in the truth of that? Can we find out whether it is true? Etc.) are addressed in well-known ways. The significance of the answers to those questions is rooted in the practices and logic of truth judging . . . , but that significance branches out beyond current practice along with the growing concept of truth. For present knowledge not only redistributes truth-values among past judgments, present knowledge also reevaluates the whole character of past practice. There is no saying in advance how this will go.[34]

On this strategy, the Tarski schema usefully expresses the deflationary spirit, not because the truth predicate is dispensable apart from a few semantic contexts, but because its applicability is too widespread and variegated to be captured by any systematic concept of truth. The *right* side of the Tarski equivalence, the many possible *s*'s that are or might be, has no systematic or coherent unity that might plausibly be captured in a theory about the left side, that is, about what it is for any '*s*' to be true.

These two approaches might seem to have different implications for whether they could be analogously extended to deflationary conceptions of knowledge. After all, 'knows' does not have a simple logical or semantic function comparable with that of 'is true', and no obvious analogue to the Tarski schema would display the redundancy or dispensability of the knowledge predicate. Yet even Horwich's approach may depend less on appeal to the Tarski schema for economically capturing the use of the truth predicate than on the intelligibility and learnability of the practices of using it which the Tarski formula schematizes (for example, attributing 'truth', relying on it, contesting it, responding to such criticism, and so on). These practices are maintained in the absence of any theoretical coherence or substantive unity

34. Fine 1986b: 149.

that would underwrite them or adjudicate the disputes they engender. Williams develops the analogy between deflationary accounts of 'truth' and of 'knowledge' in just this way:

> Naively, we might be inclined to suppose that just as in physics we study the nature of heat, so in philosophy we study the nature of truth. But once plausible deflationary views are on the table, the analogy between truth and things like heat can no longer be treated as unproblematic. . . . *Mutatis mutandis,* if we have a plausible account of the concept of knowledge, it is a further step to insist on an account of knowledge as well. A deflationary account of "know" may show how the word is embedded in a teachable and useful linguistic practice, without supposing that "being known to be true" denotes a property that groups propositions into a theoretically significant kind.[35]

In both cases, truth and knowledge, the deflationary move is a shift from thinking about a putative object that a concept could describe to thinking about the practices in which the concept is used.

This shift seems appropriate and compelling once we take seriously the dynamics of the practices of assessing, attributing, relying on, or challenging knowledge. What is at issue, of course, is not simply the use of a word but also the more far-reaching practices of assessing and assigning authority, justification, reliability, and informativeness. Both the concept 'know' and these broader concerns at issue in its use apply in significantly different contexts, for various purposes, with substantially different stakes. Particular conflicts or congruences in their application depend crucially on these variances; whether a particular application is appropriate depends on what is at stake in that case and how it is connected to other situations and contexts. The history of making and assessing claims to knowledge, the significance of those claims for other concerns and practices, and the prospects for aligning those claims with the affected agents' projects and self-understanding provide ample resources for critical adjudication of particular cases without having to appeal to a general account of the purposes and aims of science (or of knowledge more generally) to underwrite them. As Stephen Stich has argued for the cases of truth and rationality, we more appropriately appeal to locally situated concerns to assess the relevance and adequacy of any general analysis

35. Williams 1991: 113.

of 'knowledge' than to the general analysis to adjudicate particular cases.[36]

This shift of attention from scientific knowledge to the practices of attributing scientific knowledge marks a fundamental difference between my dynamic and deflationary account and the philosophical and sociological contributors to the legitimation project. The latter are driven by the idea of something underlying the practices of knowledge claims and attributions which is crucially important to get right, something that will explain those practices and either justify them or display their groundlessness. Realism, historicist and empiricist accounts of scientific rationality, and social constructivism compete to offer the best explanation of scientific knowledge and thereby to either legitimate or debunk the cultural authority often accrued by the sciences. On my account, no further or deeper explanation of scientific practices would account for their outcomes; the historically situated and contested development of the practices themselves suffices for us to understand them.

The refusal to seek further explanation or legitimation of scientific claims to knowledge does not lead to skepticism or relativism. The point of a deflationary approach is to prevent such global assessments of the "epistemic status" of scientific activities and achievements from getting off the ground by denying that "epistemic status" is a theoretically coherent target of assessment. Without such a theoretically coherent concept to survey, no grounds exist for the skeptical dismissal of scientific claims wholesale. Likewise, there can be no grounds for determining that their epistemic status is relative to the preexisting interests of particular social groups or to some other social explanans of their certification as knowledge. Attributions and contestations of knowledge are dynamically situated among many interactions among agents and with their surroundings. Which of those interactions decisively determines or justifies those attributions is not established by the supposedly socially constructed, empirically grounded, historically emergent, or causally effected nature or goal of scientific knowledge but by the ongoing dynamics of those interactions.[37] Indeed, as I shall argue in Chapter 9, philosophical or sociological interpretations of scientific practice are better understood as attempted narrative reconstructions of scientific practices from "within," continuous with the practices themselves, than as attempts to stand outside the practices and explain their (preferred) outcome.

36. Stich 1990.
37. See esp. Latour 1987: chap. 3.

A closer epistemological analogue to a dynamic, deflationary account of scientific knowledge might seem to be eliminativism. Epistemological eliminativists do not seek to explain, reduce, or otherwise account for the practices of attributing scientific knowledge. Instead, they argue that an adequate account of the domain that philosophers or sociologists of science have hoped to explain will dispense altogether with the concept of knowledge and the associated concepts of rationality and justification. The most philosophically prominent versions of eliminativism have promised to substitute the vocabularies of neurophysiology and neuropsychology for the familiar folk categories of epistemology and philosophy of science. Paul Churchland, for example, suggests that learning "is not to be explained in terms of the logical and quasi-logical relations holding of and among sets or sequences of essentially propositional states. It is to be explained in terms of the *causal* relations holding among a system of states which contain specific information as a matter of causal, nomological, and evolutionary fact."[38] Here, the notion of 'information' is a physical and quantitative concept, not a semantic one, and it marks the outcome of a causal chain rather than an interpretation. Accounts of the physical reorganization of neural pathways in response to stimuli would replace any discussion of the reasoned reorganization of beliefs in response to evidence. On such "eliminative" accounts, the folk epistemology of justification by reasoning and evidence must give way to a more scientifically adequate account of brain processes.

The details of my account of the dynamics of scientific knowledge may seem remote from Churchland's concern with brain processes, but there are distinctively social scientific correlates of epistemological eliminativism. The most familiar versions of social constructivism still offer sociological explanations of knowledge in an intentional idiom, explanations that retain the concept of knowledge as equivalent to social consensus, conceptual networks, or "the rules embodied and institutionalized in forms of life."[39] Perhaps the only thoroughgoing and self-consciously expressed social eliminativist account of scientific knowledge has been advocated by Steve Fuller: "A given science *consists of* patterns of labor organization, motivational and power structure, communication, codification, and apparatus manipulation that can be found in other, normally unrelated spheres of society, but that are made relevant to one another precisely by all these behaviors being regularly (and after a while, mutually) reinforced in a common

38. Churchland 1979: 144.
39. Collins 1992: 132.

environment."[40] Like Churchland's neural nets, these networks of social causation would supplant the fictions of folk epistemology, which mistakenly suggest that something as "uniform as a mind-set, a worldview, or even a proposition could persist through repeated transmissions in time and space."[41] Although Fuller may stand alone among social theorists in abolishing all talk of "epistemic entities" such as beliefs and reasons, softer versions of his attempted replacement of talk about knowledge and justification can be found in accounts of science for which economic, social, and demographic history take explanatory primacy over cultural and political developments.[42] Such an "ontologically permissive" social eliminativism would not rule out talk of beliefs and reasons but would make them largely epiphenomenal to the explanation of the historical development of the sciences.

Eliminativists in any form nevertheless still betray a vestige of the "epistemological realism" that deflationary approaches reject.[43] Eliminativists believe in a clearly bounded and theoretically coherent object domain that appropriately replaces the vocabulary of folk epistemology; epistemologists erred only in their description of this domain, not in their identification of it. Fuller has clearly recognized this commitment in the neurophilosophers' accounts and has criticized them for still accepting (for no good reason) traditional epistemology's faith in the epistemic autonomy of the individual organism.[44] Churchland still confines his causal substitute for knowledge and justification to those physical processes ("external" stimuli and their "internal" neuroprocessing) that take place within the boundaries demarcating a single organism from its environment. Yet Fuller's own social eliminativism in the same way presumes the traditional confines of the "social world" (the integrated object of the not-yet-unified social sciences) as the bounded domain for the causal explanation of knowledge. This unified social "metascience"[45] demarcates that part or

40. Fuller 1989: 63 (emphasis mine).

41. Fuller 1989: 4.

42. One might discern such tendencies within the recent turn toward the study of instrumentation and experimental practice within science studies, as the "material basis" for scientific knowledge comes to take explanatory primacy over the more *événementielle* disturbances of theoretical interpretation. Smocovitis (forthcoming) argues that much of what she calls "the practice industry" displays such a soft eliminativist historiographical orientation.

43. "Epistemological realism" is what Williams (1991) calls the belief that "there is a way of bringing together the genuine cases [of knowledge] into a coherent theoretical kind [so as to] make such things as 'knowledge of the external world' the objects of a distinctive form of theoretical investigation" (108–9). This commitment is not to realism within epistemology but to realism about the object domain of epistemology.

44. Fuller 1992b.

45. Fuller 1989: 63.

aspect of the world within which knowledge is located.[46] The reason that these parallel boundaries are not accidental is that eliminativist epistemologies are inherently ironical positions whose unity and integrity as domains of study are provided by their folk epistemological counterparts. What sustains the integrity of those aspects of the neural processing of human organisms which interest Churchland and those aspects of the social world which belong to Fuller's social epistemology is that they are just the features of the world which each believes necessary to supplant individual or social folk theories of knowledge.

Deflationary accounts of the dynamics of knowledge attributions make no such commitments to there being a unified object domain whose proper description would make superfluous all talk of knowledge and justification. The practices through which understanding and its recognition are distributed need have no natural bounds at all. These practices have been substantially transformed by the inventions of writing, printing, and mechanical or electronic symbol manipulation; by the substitution of instruments for perceptual discernment and manipulative precision; by the various institutions through which the credibility of human knowers was ascribed and recognized; and by many other shifts in the configuration of who and what could count as reliably or justifiably informative about the many aspects of the world. There is no reason to believe that such striking shifts in the distribution and justification of knowledge have come to an end and thus no reason to believe that the domain of knowledge (or its historically constituted subdomain of "scientific knowledge") can ever be clearly demarcated as the object of a theoretically coherent inquiry.

The deflationary turn thus in no way denigrates, dispenses with,

46. It may not be immediately obvious that Fuller has in mind such a sharply delimited social "metascience" as his epistemic explanans, because he makes frequent appeals to experimental cognitive psychology and actually includes psychology among the disciplines to be unified within an explanatory social science. But two considerations override this initial impression. First, Fuller makes primarily negative use of the results of cognitive psychology: it displays the cognitive limitations of individual knowers so as to make untenable any conception of knowledge which relies mainly on the experiential and critical capacities of individual knowers instead of social institutions and practices. Second, Fuller tacitly conceives of psychology as a bifurcated science. Its physiological aspects belong to the natural sciences (human biology), for which Fuller provides a strong constructivist interpretation in contrast to his realism about social science. Whatever remains of psychology when physiology is set aside is a social science in the strict sense. Thus, Fuller reinterprets the traditional psychological experiments that aim to study individuals' cognitive processes as laboratory studies of two-person interactions (1992b).

relativizes, or empties the concept of scientific knowledge. Raising and responding to questions about justification, reliability, credibility, and authority are pervasive and vital to the sciences and their wider practical setting. In this respect, my deflationary approach to understanding scientific knowledge is continuous both with deflationary accounts of truth and with Wartenburg's dynamic and situated conception of power. The concept of truth and the practices associated with it cannot be explained or theoretically circumscribed, not because they are unimportant, but because they are pervasive and fundamental. Likewise, Wartenburg rejects the predominant theories of social power in part because they underestimate the significance and pervasiveness of power relations and because they fail to capture the multiple, shifting, and ambiguous manifestations of power in social life. In the next chapter, I will turn to language as another philosophical model for a dynamic conception of scientific knowledge and will consider the significance of conceiving language dynamically for understanding both science and the science studies disciplines. My claim is that all three fields of interconnected concepts and practices—concerning language, knowledge, and power—are better understood as pervasive aspects of the world and not as specific domains or kinds of objects within it.

[8]

Against Representation: Davidsonian Semantics and Cultural Studies of Science

Scientific realists, instrumentalists, historical rationalists, and social constructivists have disagreed with one another about how scientific representations acquire their content and justification and especially about how representations can be connected to the objects they aim to represent. Underlying this basic conflict, however, is a largely shared conception of scientific knowledge as representational. The commitment to representationalism carries with it a recognition that knowers' access to the objects represented by scientific knowledge is a crucial problem. Even scientific realists, who claim that the best scientific representations of the world correspond to how the world is independent of us, arrive at this claim only by highly indirect arguments. Most arguments for realism accept the premise that knowers' access to objects is mediated by experience, methods of inquiry, and the theoretical representations that influence both experience and method. The realist's burden is to demonstrate, without appealing to an impossibly unmediated access to the world, that scientific theory does not distort or misrepresent things as they really are. Instrumentalists and rationalists typically settle for something less transcendent than realists aim for: they claim that scientific representations are accountable to the world as experienced or to standards of rationality or progress internal to the history of inquiry.

Social constructivists have offered stronger denials of transcendence: they take the "objects" represented by scientific claims to be an outcome of the activities that construct representations. Steve Woolgar thus summarizes the central thesis of the constructivist tradition: "We have proposed that, rather than pre-existing our efforts to 'discover' them, the objects of the natural world are constituted in virtue of representation."[1] Woolgar's work is intended as a radical, self-critical extension of this constructivist thesis. He takes constructivist sociology to partake, along with science and everyday life, in both representation and its underlying "ideology." According to Woolgar, "The *ideology of representation* [is] the set of beliefs and practices stemming from the notion that objects (meanings, motives, things) underlie or pre-exist the surface signs (documents, appearances) which give rise to them."[2] Sociologists represent science as socially constructed, but science no more transcends and corresponds to these sociological representations than nature does to scientific ones. Constructivist representations of science are thus rhetorically incoherent: "Their authors produce texts which develop and elaborate arguments for the deficiency and/or historicity of conventions of representation; but the texts themselves trade upon the same conventions."[3]

Woolgar concludes from his criticisms of the ideology of representation that science studies must direct both critical and creative attention to its own representational practices: "Without wishing to try and escape representation (which is impossible in the sense that all interpretive activities involve representation), it is nonetheless worthwhile to pursue the possibility of developing alternative forms of literary expression. The idea is that this approach might modify existing conventions and thereby provide new ways of *interrogating* representation."[4] I draw a different moral. Woolgar's polemics against naïvely representational writing and its underlying "ideology" exemplify not a new interrogation of representation but a quite old and familiar philosophical dead end. Once it is presumed that language and knowledge are representational, then we inevitably discover ourselves trapped within our representations. It is but a short step from Bishop Berkeley's ideas, which could resemble other ideas but not material objects, to Woolgar's signifiers, which could resemble other signifiers

1. Woolgar 1988c: 83.
2. Woolgar 1988c: 99.
3. Woolgar 1988c: 100.
4. Woolgar 1988c: 94.

but no object signified.[5] The familiarity of this problematic, however, seems not to have reduced its hold on the philosophical and sociological imagination. The legitimation project in which Woolgar participates along with his realist, rationalist, and empiricist protagonists depends on this representationalist presumption. This dependence is not just that the apparent gap between scientific representation and the world supposedly represented is what puts the sciences in need of philosophical legitimation. The dependence goes deeper: science can only be surveyed and "made sense of" as a whole, whether for purposes of vindication, debunking, or just explication, if scientific claims can be subsumed under a coherent kind. The presumed representational character of scientific discourse is what leverages the interpretation and assessment of scientific knowledge "as such."

In this chapter I explore the possibility of understanding the sciences while dispensing with the presumption that scientific discourse and scientific knowledge are representational. The conjunction of discourse and knowledge in the previous sentence is not accidental. Unverbalized skills of manipulation and perceptual discrimination are important for scientific practice, but scientific understanding is still undoubtedly permeated by language. Moreover, the usually unargued presumption that language works representationally has strongly encouraged representationalist accounts of scientific knowledge. If scientific theories and other hypotheses are meaningful as representations of possible states of affairs in the world, then it seems quite natural to think of their epistemic standing in terms of whether these representations correspond to the objects they represent. Yet within the philosophy of language, in both the Anglo-American and West European styles, there are now well-developed, if still highly contested, alternatives to representationalist semantics. In this context, it is (or should be) surprising that philosophy and sociology of science still proceed mostly as if representationalist conceptions of language and knowledge were the only game in town.

5. The target of my criticism is Woolgar's conception of representation as omnipresent and not his particular explorations of various practices of inquiry, which may not require such representationalist underwriting. Indeed, while Woolgar's general criticism of the ideology of representation resembles the all-encompassing epistemological skepticism about the external world articulated by Descartes and Berkeley, Woolgar's own skeptical practice is often closer to that exemplified by Sextus Empiricus or Montaigne, for whom skepticism is above all a moral practice aimed at avoiding philosophical dogmatism in one's life. Woolgar's pleas (1991, 1992), to avoid "positionism" could be plausibly regarded as an articulation of such a "classical" skepticism in lieu of the claims I criticize here.

This chapter can thus also be understood as a reflection on the possible significance for the philosophy of science of taking seriously the criticisms of representationalism. My own presentation draws on the work of Donald Davidson and his followers, but other antirepresentationalist approaches could have served instead.[6] I make no pretense here of defending my own broadly Davidsonian commitments against other ways of understanding language and thought; that project would belong to a different book. My more modest goal is to present an attractive alternative to the representationalism that most science studies have taken for granted and to suggest how such an account of understanding and communication might contribute to a different approach to understanding scientific practices.

COMMUNICATION WITHOUT REPRESENTATION

Woolgar's reflexivist criticism of the constructivist project offers a useful way to introduce the importance of nonrepresentationalist semantics for science studies. Woolgar presents reflexive constructivism as a radical extension of the Strong Programme in sociology of science. The constructivist pioneers of Edinburgh and Bath allegedly fell short of a thoroughgoing relativism because they failed to be fully constructivist and relativist about their own rhetorical productions.[7] Woolgar, by contrast, takes the reflexive application of constructivist sociology to be "the hardest possible case" for constructivism and therefore the most interesting and revealing. From another perspective, however, Woolgar's "interrogation" of representation looks remarkably timid. Woolgar insists that we can never get beyond representations to encounter the objects they supposedly represent. We have many versions, but no world of which they are versions. But Woolgar evinces no comparable doubts that we could ever get *to* representations in the first place. How does a string of perceived marks or sounds ever count as a meaningful version? How do we ever represent the world *as* being one way rather than another?

6. Davidson is often interpreted as a less radical critic of representationalism than my interpretation of his project would suggest. My reading is strongly influenced by Ramberg 1989; Wheeler 1986, 1989, 1991; and Rorty 1989, 1991. Wheeler's essays develop specific connections between Davidson's project and the seemingly very different criticism of representationalism developed by Derrida (1973, 1976, 1988). My earlier discussion in Rouse 1987b suggests how the Davidsonian tradition might also connect to Martin Heidegger's hermeneutical philosophy.

7. Woolgar (1992) develops the contrast between an epistemologically "radical" reflexivity and the timid, not-yet-relativist work of the Edinburgh and Bath schools.

My point in raising this question is not to encourage general doubts about whether thoughts and utterances are ever meaningful. Rather, I want to encourage doubt about Woolgar's presumption that representations (that is, their meaning or content) are more accessible to us than the things they supposedly represent. If there is no magic language through which we can unerringly reach out directly to its referents, why should we think there is nevertheless a language that magically enables us to reach out directly to its sense or representational content?[8] The presumption that we can know what we mean, or what our verbal performances say, more readily than we can know the objects those sayings are about is a Cartesian legacy, a linguistic variation on Descartes's insistence that we have a direct and privileged access to the contents of thought which we lack toward the "external" world. Skepticism and epistemological relativism, along with their alleged refutations, depend on this primacy of the mental or the meant for their intelligibility, their significance, and their argumentative support.[9]

Davidsonians deny that the meaning or content of linguistic performances is more readily accessible than the world they refer to.[10] Marks on the page, blips on the screen, vibrations in the air, or processes in the brain are events like any other. Uttering sentences and responding to them are among the many things people do, and the environment in which human beings make their way includes patterns of utterances (and the patterns of other activity which can be interestingly correlated with utterances). Making sense of these utterances (including making our own utterances sensibly) is part of our overall activities of making sense of and coping with the world we live in. Nothing mediates our interaction with the world, not experience, thought, language, meaning, or representation. Talk and perception are just further interaction, not a medium through which interaction is filtered.

It is easy to misunderstand drastically this denial of mediation. The claim is not that the world or its objects are directly and immediately

8. I take the perspicuous phrase "magic language" from Wheeler 1991.

9. Williams (1991) has been especially clear in linking skepticism and the epistemological project of assessing knowledge as a whole to this presumption of the priority of experience and meaning.

10. I use "Davidsonian" advisedly. My intent is not to explicate Donald Davidson's philosophical work but to display a philosophical approach which I put forward on its own merits but which I take Davidson most clearly and originally to exemplify (with occasional backsliding). Perhaps in the end Davidson is no Davidsonian in this sense, but that is of little moment to my concerns.

present; rather, nothing is directly present, not even experience, thought, or meaning.[11] Understanding any one thing (an object, an utterance, some other perceived event, or a pattern of objects and events) requires understanding many others. A noise cannot be an utterance or a movement an action unless they belong to a larger pattern of utterances and actions within which they are intelligible. This Davidsonian commitment is often described as holism, in contrast to atomistic views for which individual words or sentences are meaningful by themselves (perhaps in virtue of their referential relations to objects or facts) and more complex patterns of meaning are assembled from these previously meaningful parts. But the term 'holism' should not be misunderstood to imply a totality of meanings or beliefs which would enable us to speak of a language, a belief system, or a culture as a whole. The commitment is only to the constitutive interdependence of many meanings and beliefs, such that revising any one interpretation will require adjustments elsewhere.[12]

The denial that language works representationally is also seriously misunderstood if it is thought to imply prelinguistic access to the world. For Davidsonians, language permeates all understanding, and there is no way to get back "outside" of language.[13] Davidson's semantics is directed toward the interpretation of occasioned utterances. The theory presumes that one already speaks a language, and it takes interpretation within an unanalyzed home language to be the most fundamental level of linguistic understanding. One's own language does not thereby become a privileged frame of reference, for it no more supplies a fixed "frame" than does any other way of speaking. Expressions in one's own language can be interpreted in turn, but only by construing them within a language that must remain unexamined for that occasion.

For Davidsonians, occasioned utterances are interpreted on the basis of what is said when. The circumstances in which the utterance is made, together with what has been said and done in other circum-

11. Derrida (1973) offers a more striking and seemingly paradoxical formulation of a similar claim: "There never was any 'perception' " (103).

12. Wheeler (1986) makes this point in passing while noting the far-reaching implications of Davidson's doctrine of an apparently modest indeterminacy of meaning: "If we do not make the simplifying assumption that a person's language or theory *at a time* is a unified whole, then areas of our own language are indeterminate relative to other areas of our own language" (490).

13. Obviously, Davidsonians recognize that people are not born already speaking or understanding in language. But they would insist that first acquiring language is a very different matter from learning an additional natural language and even more so that one cannot abstract from our linguistic competence so as to isolate some prelinguistic "component" of our interaction with the world.

stances, provide the only evidence to which interpretation is accountable. Because meaning is not directly accessible, we can arrive at it only by tacking back and forth between sentences and circumstances, as we understand them. Any one utterance can be interpreted only against a background of other beliefs. We have no basis for attributing such beliefs, however, except to presume that those we interpret mostly believe what is true about our shared circumstances. We have no access to what is true other than the sentences we take to be true in our home language. So a necessary constraint on interpretation is that we take those we interpret to agree with us about most of the many homely truths about the world. Does this mean that everyone really shares a large stock of common beliefs? No, it means that there is no agreement or disagreement, indeed, no belief at all, except against a background of postulated agreement. Without such postulation, no basis exists for assigning *any* interpretation to someone's utterances (including to one's own; for a Davidsonian, there is no privileged access to any meaning "behind" one's own utterances, which are therefore interpretable texts like any other).

The Davidsonian view that interpretation posits massive agreement in beliefs rather than discovering it may seem initially counterintuitive. "Beliefs" may seem to be already lurking in people's head prior to interpretation. But for a Davidsonian, "belief" is a theoretical concept, one applied only as part of an interpretation: to make sense of a particular utterance or other action, the interpreter must postulate a wide range of sentences the speaker "holds" to be true at the same time. Of course, we continually treat ourselves in the same way when we make our own utterances. For what we say to be intelligible even to ourselves, we must postulate a wide range of other sentences that we "believe," that is, hold true at the same time. To speak is always simultaneously to interpret oneself as a speaker, in the same way we interpret others. The apparent preexistence of belief would then be merely the outcome of the ongoing interpretation of ourselves which is realized whenever we speak or think.

Davidsonians describe the interpretations through which speakers understand one another as truth theories. To make sense of why truth plays such a fundamental role, one must first understand how truth might be completely divorced from the concept of representation. Truth is too often construed epistemologically as a relation of correspondence between sentence meaning and facts or of coherence among meaningful sentences. Truth is then analyzed in terms of what could *make* those already meaningful sentences true about the world. For Davidsonians, this is topsy-turvy. Sentences do not "have" meaning except on the basis of what is (or would be) said when. Thus, establishing which sentences a speaker holds true provides the only basis for interpreting what those sentences could mean. Getting the speaker's sentence constructions to come out mostly true (by mapping them onto sentences *we* hold true) is what enables an interpretation of the speaker's utterances to count as satisfactory. There simply is no meaning prior to such provisional assignments of truth-values. For Davidsonians, truth is thus a purely semantic concept that enables sentences to be interpreted by mapping them onto the sentences the interpreter would accept on the same occasions. Nothing "makes" sentences true, and 'truth' does not name any relation, certainly not a relation between sentences and anything nonlinguistic. Failure to appreciate this point often results from a misunderstanding of the Davidsonian principle of charity. The principle is that interpretation can proceed only by positing massive agreement between speaker and interpreter, or, better, by taking speakers to speak and believe mostly truths. The principle is misunderstood when it is taken either as a heuristic maxim or as a (highly counterintuitive) epistemological claim. Both misreadings still presuppose a representationalism that would fix meaning independent of worldly circumstances. Ramberg nicely characterizes this common mistake:

> Just as we have no choice, if we want to make sense of what others say, but to regard them as on the whole speaking the truth, so we have no choice but to regard *ourselves* as largely speakers of truth. The principle of charity is no overestimation of our cognitive and communicative capacities. Massive error, just like massive deception, presupposes an epistemic, representational relation between language and the world, and a corresponding reification of the meanings of words and sentences. . . . The unfortunately named principle of charity appears charitable only when we forget that the insight an interpreter gleans when isolating the causal relations between the world and a speaker's beliefs is *semantic,* not epistemic, and that these causal relations are not available for justificatory purposes.[14]

No one has ever explicitly constructed a complete truth theory for a speaker or a natural language and such a construction should not be anticipated. But truth theories themselves do not figure crucially

14. Ramberg 1989: 99.

in Davidsonian semantics. What interests Davidson is interpretation, which he models on the ongoing activity of constructing truth theories.[15] The theories being constructed map utterances in the speaker's vernacular onto utterances in the unanalyzed home language.[16] For Davidsonians, the primary phenomenon to be understood is linguistic interaction, or *communication.* What is most striking about the Davidsonian approach is that it accounts for such communication without reifying meanings, reference, or structured languages. We do not have to postulate meanings already present to speakers which are then passed on to an audience; nor do we need referential connections between words and things to fix meaning; nor is it necessary to assume that speakers and their interlocutors must share a conventionally accepted language. The entire apparatus of representation is dispensed with, and communication is modeled as a dynamic interaction among speakers in an environment of words and things.

The theories that interpreters construct are not systematic ones for a whole language or idiolect but "passing theories," temporary constructions that are continually being revised to make sense of speakers' ongoing utterances.[17] These theories make sense of what people say and believe by adjusting the interpretation in the light of changing circumstances and a growing stock of utterances to interpret. Davidson takes as a datum to be accounted for that ordinary, successful communication is still rife with malaprops; jokes; indexicals; unusual word choices; innovations; mistakes in spelling, pronun-

15. The importance of this point for understanding Davidsonian semantics has been brought forcefully to my attention by Ramberg 1989: 78–81.

16. Considerable technical apparatus underlies Davidson's notion of a truth theory. The central idea, which inverts Tarski's 1944 project of analyzing the concept of truth, is to postulate a finite group of axioms which could generate an infinite number of theorems displaying the truth conditions of individual sentences. The theorems establish material equivalences between the *truth* of the sentences in the language being analyzed and corresponding sentences in the home language (metalanguage) in which the analysis is conducted. The theorems thus take the form, " 'S' is true if and only if *p*." Material equivalences, of course, make no reference to intensional features of the equivalent sentences; some further constraints are needed to connect the truth of "snow is white" to snow being white rather than to grass being green or the sun rising in the east. These constraints are provided by the same mechanism that generates the infinity of possible theorems from a finite number of axioms: the theorems are constructed by recursive application of rules for composing sentences from their parts. "Snow is white" is connected to the color of snow by the constraint that the theory must economically account for other sentences that employ "snow" and "white" in different contexts, consistent with the principle of charity. That is, the truth-values of most of the sentences that the speaker accepts must map onto sentences to which we would assign the same truth-value. Thus, although sentences can be construed only holistically, the overall interpretation in a truth theory works by taking them to be truth functionally composed of recombinable parts.

17. The notion of a "passing theory" is introduced in Davidson 1986.

ciation or grammar; and many other occasional and nonstandard uses of words. Of course, speakers enter conversation with many expectations about what people are likely to say and how words will usually connect together. The amalgamation of those expectations as a "prior theory" certainly contributes to speakers' understanding (and misunderstanding) of one another. "Prior theory" is regularly overridden, however, as successful communication requires only a less systematic or stringent convergence of the interlocutors' passing theories.

Thus interpretation is always dynamic according to Davidson and this dynamism dispels the illusion that "meanings" (including meaningful actions) constitute a distinct or coherent domain of objects.[18] Indeed, meanings are not objects at all. There are only utterances, actions, and the dispositions to further actions and utterances which interpreters posit in making sense of one another (including making sense of themselves). This claim is strongly antireductionist, however. Instead of reducing meaningful utterances to something meaningless (for example, behavior or syntactical rules), Davidsonians understand the entire world we inhabit to be linguistically configured. As I noted earlier, Davidsonians accept language as an irreducible and inseparable dimension of all understanding.

The irreducibility of language is coupled with Davidsonians' rejection of the analytic/synthetic distinction. Without a principled distinction between meaning and fact, there can be no clear line between understanding what people say and understanding other aspects of the world. Samuel Wheeler summarized this consequence of understanding language nonrepresentationally:

> Without a magic language whose terms carry meanings by their very nature, the determination of what sentences mean and what is true, that is, what the facts are, rests on a single kind of data, what people say when. Thus, there is no separating learning a language from learning about the world. . . . If this is true, then there is likewise no difference between "contingencies" of what we say and might have said and contingencies about what the world is or could have been. So, changing language is continuous with changing the facts; and changing the facts is continuous with reevaluation.[19]

18. The idea that meanings are a special kind of object has been an influential basis for conceiving of the human sciences as methodologically distinct from the natural sciences. For extended arguments against using this idea to establish such distinctions, see Rouse 1987b: chap. 6; Rouse 1991b; and for developing distinctively Davidsonian arguments against the distinction, esp. Rouse 1991a.

19. Wheeler 1991: 201.

These conclusions may seem startling when we tacitly presume that meanings are fixed independently of how the world is. How could the world change because of what people say? But the Davidsonian point is that what people's utterances could possibly mean depends on what is true and that what is true can only be said in a language. There is a world prior to linguistic understanding but no meanings and no way the world is.

These claims appear more natural and obvious once we recognize that utterances and other interpretable actions are prominent features of the world we inhabit and try to understand. The patterns of people's utterances do not merely respond to preexisting features of the world but also introduce new features and discriminations. For example, Wittgenstein points out that the various behaviors and circumstances that count as "expectings" do not have some unifying properties (or even family resemblances) prior to the practice of using the term 'expect' to talk about them. This linguistic practice is an ineliminable part of the phenomenon of expecting. Yet 'expect' does not already have a meaning apart from the ongoing practices of using it in various situations: the signification and its having significance happen together. Thus, Wittgenstein concludes that "it is in language that an expectation and its fulfillment make contact."[20] The world that we inhabit and understand is suffused with linguistic practices. Understanding those practices requires knowing many other things about the world, while much of the world is incomprehensible without understanding those linguistic practices. To undercut distinctions between meanings and facts is to deny that these aspects of our understanding can be usefully disentangled.

Metaphors and other tropes become especially prominent on such accounts of language as nonrepresentational. Tropes have typically seemed derivative when language was understood to be representational. They could only supplement or extend literal or representational content by representing in a different way, "metaphorically." Davidsonians insist instead that tropes are crucial to the ways we use language but do not represent or express anything. They depend on familiar ways of using words (someone who does not understand the way the words are more ordinarily used will not normally "get" a metaphor or other trope), but they do something different from more humdrum uses. When tropes work, they "turn" us, cause us to

20. Wittgenstein 1953: par. 445. Wittgenstein's discussion of expecting extends through paragraphs 444–53 and 569–86. I discuss these passages more extensively on p. 222.

attend to or respond to things differently.[21] Typically, truth-theoretical interpretation signals a trope when an utterance comes out as obviously false (or, in some cases, trivially true) on the standard expectations the interpreter brings to the interchange. Either one must significantly revise one's interpretations or accept that the speaker is doing something other than saying what's what. Of course, not all such prima facie false statements are tropes; their initial manifestations in a truth theory do not distinguish them from errors, misunderstandings, previously unrecognized truths, or jokes. Tropes stand out from other unexpected uses of words in the responses they evoke: they amuse, provoke, associate, resound, and so forth. Davidson's point is that no intermediate "metaphorical content" is needed to accomplish these diverse roles. Familiarity with what is typically said when and an imaginative response is enough.

> Understanding a metaphor is as much a creative endeavour as making a metaphor, and as little guided by rules. [This remark] does not, except in matters of degree, distinguish metaphor from more routine linguistic transactions: all communication by speech assumes the interplay of inventive construction and inventive construal. What metaphor adds to the ordinary is an achievement that uses no semantic resources beyond the resources on which the ordinary depends.[22]

The importance of metaphor on the Davidsonian approach is that it both expands the range of what can be done with words and transforms the possibilities for ordinary truth saying. Metaphors can slip into more ordinary uses, doing so in ways that rearrange other truth-values. If trees come to have rights "literally," that is, as an unproblematically accepted truth, then either rights will matter less or trees will be treated differently (or perhaps the change will be accommodated by more subtle adjustments). This recognition that changes in language are consequential and shift truth-values in ways that cannot be neatly parceled out as changes in meaning or changes in fact has significant political consequences. These consequences are often misunderstood by enthusiasts as well as by critics and are reassimilated to a social relativism. Thus, before we can adequately consider the political significance of a Davidsonian approach to language, we must clear the ground of some errors.

21. Donna Haraway and Betty Smocovitis indicated to me in conversation that a *tropos* might ordinarily be translated as a "turn."
22. Davidson 1984: 245.

A good example of the kind of interpretation we should reject is Barry Allen's conclusion, from broadly Davidsonian premises, that "there is nothing more to the truth (the being-true) of the occasional truth than the historical fact that on that occasion what is said . . . passes for true."[23] Allen's conclusion is mistaken on multiple grounds. First, for Davidsonians, the concept of truth cannot be analyzed at all; hence, it cannot be analyzed as "passing for true" or "accepted by a community." But this reading also misunderstands the significance of interpretive charity and fails to recognize adequately that Davidsonian interpretation is dynamic. On the former point, what passes for true is where interpretation must begin, not end. Once we know which sentences a speaker (or community of speakers) accepts, we must still make sense of what those sentences say, which can be done only by a holistic interpretation that makes most of the speaker's beliefs come out true.[24] We thus cannot know *what* it is that passes for true on an occasion, except by reference to what is true.

This insistence on the primacy of truth does not merely replace what passes for true for the speaker (or her community) with what passes for true for the interpreter. A Davidsonian truth theory interprets an utterance via its truth conditions, the circumstances in which it would be true. These circumstances must always be described, so to that extent, as Richard Rorty puts it, Davidsonians "explain *true* in terms of *language I know*."[25] But what is decisive are the truth conditions themselves and not the particular language chosen as an interpretive metalanguage. The metalinguistic description of a sentence's truth conditions is itself open to interpretation and criticism, which can only proceed in turn by another holistic specification of truth conditions.[26] Indeed, any such actual metalinguistic interpretation is

23. Allen 1993: 6; Allen's claim receives extensive critical discussion in Okrent 1993.

24. It is crucial to remember that a Davidsonian truth theory must account for not only the truth-values assigned to individual sentences but also the truth-values assigned to the infinite variety of other sentences that might be constructed from its components. The compositionality of sentences thus places a strong constraint on truth-theoretical interpretation.

25. Rorty 1991: 157; Rorty is actually quoting Jonathan Bennett, but Bennett takes this description of Davidson's project as a criticism, whereas Rorty extols its virtues.

26. At some point, one is usually reduced to homophonic interpretation: 'Free quarks have a charge of 1/3 *e*' is true if and only if free quarks have a charge of 1/3 *e*. What the apparent banality of homophonic interpretation shows, not surprising if one thinks about it, is that one's understanding of what is being said is likely to be no better than one's grasp of the matters under discussion. On occasion, of course, the homophonic interpretation can be genuinely informative, typically when one's initial construal suggested a different version. Further discussion may indicate that someone's earlier utterance was not a joke, a slip of the tongue, or a provocation but that the speaker actually meant what she said (and that you were speaking the same language).

part of the dynamic process of interpretation. Interpretation is always provisional, pending further interaction with speakers and circumstances. Thus, to repeat, identifying truth with acceptance, or with "passing for true," gets things exactly backward. Understanding what we and others accept requires ongoing interaction with the world and what is true about it.

Once we recognize the significance of interpretive charity, it may seem that a Davidsonian semantics undercuts any constitutive political significance to language.[27] This impression is also misleading, however. It arises because of the presumption that truth is a closed concept, that is, that truth-values already exist for all possible sentences, even for those as yet unthought, unsaid, or unknown. But natural language is much looser, the world more unsettled, and truth more open than this presumption would allow. As Wheeler has noted:

> The philosophers' notion of "true" and "false" seems to (almost) fit when the practices are (almost) completely fixed, so that it is already (almost) laid out in advance, for every possible object and kind of object, whether that object is in the extension of the term. . . . We could expect good application of "true" and "false" in the case of totalities that we can figure out in advance, like the numbers and the sets of grammatical sentences in first-order quantification theory. There, we can prepare for every eventuality, in those terms. . . . But in natural languages, the very items that are to be elements of sets are not given by the nature of things. There is no manifold prior to conceptualization, that is, prior to language.[28]

Davidsonians allow no place for alternative conceptual schemes that would make for general indeterminacy or incommensurability, because there is no manifold to be schematized differently. The only basis for saying that languages are the same or different is interpretation, which presumes widespread agreement. Yet disagreement is not thereby abolished but only localized, and it might conceivably be localized differently consistent with all the evidence.

The significance of such marginal indeterminacy is expanded once we recognize that the locations of the margins are not fixed in advance of interpretation. As Davidson notes, it is a "mistake to suppose

27. Okrent 1993 offers an especially cogent account of why Davidsonian pragmatism cannot reduce truth to passing for true and seems inclined to draw this conclusion about the argument's political significance.

28. Wheeler 1991: 211.

there is a unique language to which a given utterance belongs. We can without paradox take that utterance to belong to one or another language, provided we make allowance for a shift in other parts of our total theory of a person."[29] A language, in this sense, is an interpretive reification, consisting of the domain of utterances for which one constructs a single truth theory. Wheeler reminds us that it is only "a simplifying assumption that a person's language or theory *at a time* is a unified whole, [without which] areas of our own language are indeterminate relative to other areas of our own language."[30] Comparable assumptions must be made to include a speaker's utterances over time or the utterances of more than one speaker within a single truth theory. Recognizing these considerations, Bjorn Ramberg concludes that "what is indeterminate is not what a sentence of a language means, *but what language is being spoken.*[31]

The political significance of these indeterminacies arises from the role of interpreters in specifying the scope of the interpretation, that is, what language is being spoken. For Davidsonians, convention is not crucial to linguistic meaning, and idiosyncratic speech is not uninterpretable, so the political issue is not that some things cannot be said within the language being spoken. The issue instead is the role of power relations in determining how various speech acts are situated among other uses of words and encounters with things. Should an unexpected utterance be assimilated to familiar uses or taken as a sign of linguistic divergence?[32] If the latter, does it signal an unexpected fact, a new way of putting things, or just a misuse of words? For Davidsonians, these differences are not intrinsic to the utterance in isolation but depend on the circumstances in which it is situated. Moreover, the circumstances in question are not yet fully settled: part of what matters is how speakers respond to what was said, and what happens subsequently.

When (possible) changes in language are at issue, metaphor, metonymy, and other tropes may play an especially significant role. Tropes have many uses quite removed from truth saying, but they can also be occasions for reconfiguring the connections among things, utterances, and other practices. Wheeler states:

29. Davidson 1984: 239–40.
30. Wheeler 1986: 490.
31. Ramberg 1989: 90.
32. Strictly speaking, of course, an utterance being unfamiliar or unexpected is irrelevant; the most striking semantic divergences often arise from familiar constructions being redirected in ways that may not be readily apparent.

> Sometimes our intention in saying such things as "Criminals are victims of their environment," "Fetuses are persons," or "Wildernesses have rights" is to *urge* that the statement be true. This urging, especially when the topic matters, is not just observing the obvious, but it is not really distinguishable from literal predication, either. . . . The point of the struggle is that a lot hangs on this categorization. If you are a victim, you have been attacked unjustly, there is an attacker, and compensation is owed.[33]

Given that practices and things are connected in multiple ways, shifts in meaning and the drift of tropes into ordinary truths can thus have significant consequences. Making one adjustment is likely to require multiple adjustments in what is said and done elsewhere. The point is not that language determines the facts or compels some meanings instead of others. Language is no more a privileged means of political control than it is a neutral medium in which political struggles can be reported and assessed. Language neither comes between us and the world nor stands outside of the world as a field in which its objects can be re-presented. Utterances and the patterns they can form are both an inextricable part of the environment people cope with and an indispensable part of coping with it. Understanding what people are saying and understanding what is the case are part of the same ongoing activity of making sense of ourselves and the world.

SCIENCE STUDIES AFTER DAVIDSON

Appealing to philosophy of language to resolve issues for the philosophy of science has yielded repeated disappointments. Empiricist theories of meaning, postpositivist theories of meaning change, and causal theories of reference have all come to grief after initially promising a more secure ground for philosophical interpretations and legitimations of scientific knowledge. The importance of Davidsonian philosophy of language for science studies is not to have succeeded where these earlier approaches failed. Reflection on Davidson's work instead helps to show why these failures were inevitable and to guard against their repetition in new or hidden guises. We can never expect to understand scientific knowledge by understanding the workings of language, because their "working" is too closely intertwined. Learning

33. Wheeler 1991: 213.

our way about the world and learning our way about our language are part of the same process.

Taking the Davidsonian project seriously would have both critical and constructive consequences for subsequent work in science studies. Critically, Davidson's work focuses and extends the suspicions I have already tried to raise against the legitimation project and its underlying representationalist epistemology. In particular, once we see that representationalism is a contentious assumption, realism, empiricism, historical rationalism, and social constructivism all appear to be vestiges of the misguided attempt to "illuminate the nature of science by an examination of alleged necessities of language."[34] Constructively, a Davidsonian critique of representationalism encourages a recognition of the multiplicity and heterogeneity of the connections between scientific practices and other features and aspects of the world. This recognition accords with an alternative approach to science studies as cultural studies of science. Cultural studies emphasize a critical engagement with discursive practices. There is no way to get outside or behind language in order to explain discursive practices in other terms. Yet language does not constitute a self-enclosed system of meanings that can be understood apart from ongoing practical engagement with the world. For both Davidsonian semantics and cultural studies, meaning is best understood as an open and dynamic engagement with the world. On both accounts, scientific practice and cultural studies of science participate in that engagement and do not simply disclose already determinate meanings, whether of the way the world is or of the nature and aim of science.

My account of the critical significance of Davidsonian semantics for science studies may seem constrained by the limits of my presentation of Davidson. I have described the Davidsonian project as an attractive and coherent program for understanding linguistic communication without reifying meaning or representation, but I have not attempted to defend it systematically against alternative programs in the philosophy of language and mind. Even this limited exposition may nevertheless have important critical implications for the dominant interpretive programs in philosophy and sociology of science, because their commitment to epistemic representationalism has been mostly unargued. Realists, rationalists, empiricists, and social constructivists

34. Shapere 1984: 383; as I argued in Chapter 1, Shapere is mistaken in thinking that his own project has abandoned this commitment to the philosophical priority of the "necessities of language."

have generally presumed that knowers have relatively straightforward access to the content of their thoughts and experiences, but that epistemic access to the objects of those thoughts and experiences is more problematic. The differences among their responses to the felt need for a general legitimation of scientific knowledge can be understood largely as disagreements over how best to characterize and resolve this alleged difference in epistemic access. At the least, the availability of a serious and sustained alternative to this representationalist presumption means that their shared commitment to it is in need of argument. Furthermore, a failure to establish such arguments would be devastating. Davidsonian and other nonrepresentationalist positions do not just challenge the viability of their answers to the question of how scientific knowledge is legitimately related to the world; they impugn the seriousness and importance of asking or answering that question at all.

The Davidsonian challenge is strengthened by more detailed consideration of its import for each of the specific traditions that take up the legitimation project. In the case of scientific realism, the availability of a nonrepresentationalist semantics demolishes much of what has made realism seem attractive despite its repeated argumentative failures. Realism's intuitive appeal has two powerful sources: the manifest familiarity of ordinary physical objects (such objects may seem more readily accessible than the experiences or social relations that antirealists propose as mediations), and the lack of any sharp distinction in everyday practices between the observable and the unobservable or the social and the natural. Davidsonian semantics appeals to these same intuitions, however, while denying the characteristic doctrines of scientific realism (the mind- and language-independence of the world and the explanatory power of truth and reference). The realist doctrines are thus severed from the most persistent and plausible reasons for believing them. Moreover, the Davidsonian criticism isolates precisely what is objectionable about scientific realism: not its acceptance of the truth of scientific theories or the existence of theoretical entities but its commitment to semantic realism. Realists believe that scientific theories say something definite *apart* from their many practical interconnections with the world and that the world has a preferred description prior to those same interactions with language users. Neither semantic doctrine shares the spontaneous appeal often conveyed by the term 'realism.'

Davidsonians also challenge what is characteristically appealing about empiricist and historical rationalist interpretations of the sciences: the supposed proximity and accessibility of observation or methods of inquiry. Both traditions trade on the presumption that our own experiences or procedures are more directly familiar than those things "beyond" us which we can only describe. But if their relative accessibility cannot be presumed, then the experiences or methods that legitimate scientific knowledge must be specified. Any serious attempt to describe our experiences or methods, of course, can only confirm the fundamental Davidsonian insight that for the understanding there is nothing better than words and nothing other than further descriptions that "makes" our descriptions of the world true or false, justified or unwarranted.

The Davidsonian challenge to social constructivism further exposes its unwanted affinities to the classical philosophical doctrines about knowledge as mental representation. Social constructivists have often chided philosophers of science for a naîve allegiance to notions of mind, reason, and an objective world, one that allegedly ignores the importance of human agency and the constitutive role of social interaction in the creation and establishment of meaning. But we need to ask how "social" interaction is being conceived here. Social constructivism has emerged from a tradition in social theory and the philosophy of social science that identifies the distinctively social domain as composed of rules, institutions, and forms of life.[35] This tradition originally aimed to preserve the autonomy of social science from physicalist or behaviorist reduction. Social constructivism simply moves beyond the autonomy of the social "world" from the natural world and to its epistemic and metaphysical priority over nature. From a Davidsonian perspective, however, the reification of meaning or representational content is a fundamental bond between social constructivism and the epistemological tradition that is needed to confer significance on their differences. The balance of activity and receptivity in the (individual or social) knowing subject and the relative priority of cognitive or social analyses of meaning transactions are subordinate questions. Thus, for Davidsonians, Woolgar is right on the mark in seeing the question of how to understand representation as the central theme of the disputes between social constructivists and more traditional epistemologists. Whereas Woolgar sees an important issue finally emerging clearly, however, Davidsonians see the entire debate exposed as a consequence of false premises.

35. Two classic sources for this move are Winch 1958 and Berger and Luckmann 1966. The roots of the former are in the later Wittgenstein; the latter, in Husserl read through Alfred Schutz.

If we accept the Davidsonian program, the debates over how to construe and legitimate scientific representation are otiose. Do Davidsonians then have an appealing alternative research program for understanding and critically engaging the sciences? It is crucial not to overstate the importance of the philosophy of language for science studies. By itself, a Davidsonian philosophy of language entails very little about how to proceed in understanding scientific practices and accomplishments. Nevertheless, I see an informative convergence between a Davidsonian understanding of language and a promising program for science studies as cultural studies. To see how we might get from Davidsonian semantics to cultural studies of science, I would like to recount briefly Nancy Cartwright's critical discussion of what she calls "fundamentalism" about scientific knowledge. In her first book, *How the Laws of Physics Lie,* Cartwright argued that the fundamental laws of physics are false: they acquire generality and explanatory power only through abstractions and idealizations that preclude their accurate description of any real objects or situations in the world.[36] Without denying the value of the explanatory power and calculative economy embedded in fundamental laws, Cartwright nevertheless identified what physics can say truthfully about the world with the less systematic or powerful phenomenological laws that connect fundamental idealizations to concrete situations.

Cartwright now believes that she misidentified the target of her criticisms. The problem lies not with the truth of physical laws but with their scope.[37] Like most philosophers of science, Cartwright originally understood fundamental laws to have unlimited scope; the laws themselves specified an unlimited domain of application which, she then argued, far exceeded what they could accurately represent. Cartwright now suggests that the mistake was in presuming that laws specify an unlimited domain of application. This commitment to unlimited scope she calls "fundamentalism." Whereas Cartwright herself now takes "the point of scientific activity to be building models that get in . . . all and only those circumstances that the laws in question accurately describe, fundamentalists want more. They want laws; they want true laws; but most of all, they want their favorite laws to be in force everywhere."[38] Cartwright understands her criticism of funda-

36. Cartwright 1983.

37. In Rouse 1987b: 148–52, I argued similarly that Cartwright's 1983 arguments were compatible with the truth of fundamental laws in physics, if one made suitable adjustments in the interpretation of the law and its scope.

38. Cartwright 1994.

mentalism to be a metaphysical claim, but it is perhaps better understood to be semantic. The metaphysical claim would be that reality is "a patchwork of laws": the regularities the world exhibits are only *ceteris paribus* laws, or manifestations of tendencies of things, not strict universal laws.[39] The semantic claim is that interpreting laws and theories is not distinguishable from articulating models that handle concrete situations in the terms of the theories: theories apply only where reasonably accurate models can be found.

Cartwright's project becomes clearer if we see her as trying to avoid both fundamentalism (laws apply everywhere) and empiricist reductionism (the content of laws is restricted to their observable consequences). Both fundamentalists and reductionists think laws and theories already have a definite domain of application fixed by their meaning. In the one case, that meaning is taken to be expansive; in the other, quite constricted. Indeed, if one is attentive to the theological overtones of Cartwright's chosen term for her antagonists, with its connotations of dogmatically "literal" interpretation of divine law (which in this case, of course, is natural law), then reductionism is also a kind of fundamentalism. Reductionists and more expansive fundamentalists differ only in their preferred hermeneutics of natural law.

For Cartwright, by contrast, theories and laws do not come ready-made with a definite meaning or a preferred interpretation. Theories are meaningful only through the construction of models that connect their terms in specific ways to families of concrete situations in the world. For example, the seventeenth-and eighteenth-century development of the standard models of classical dynamics (for example, for variously forced or damped harmonic oscillators) was simultaneously an interpretation of the theory and a discovery about the behavior of things in the world. Similarly, Darwin's theory that species originate from the operation of natural selection on chance variations in individual traits is further articulated by models of sexual selection and kin selection. What a theory says about the world depends on how various situations are or could be modeled in its terms.[40] The

39. More radically, the metaphysical claim might be that the very notion of law is misconceived, a last vestige of natural theology. Such a view would admit both causes (singular events that cause other singular events) and localized regularities, but no general laws.

40. In the expanded sense of "fundamentalism" that equates it with the assertion of fixed meaning according to a preferred interpretive strategy, Cartwright's own position could also be susceptible to a quasi-fundamentalist reading. If one thought that the statement of a law or theory strictly constrained in advance what could count as a model of it, then the meaning and scope of the theory would be determined in advance by the statement of the theory

content of the theory is not limited to the set of models already articulated, but nothing has already determined whether and how it applies to situations not yet specified in that way. The acquisition of new models for a theory thus neither changes its meaning (as reductionists might have it) nor merely spells out explicitly what it had already meant implicitly. Indeed, we are better off thinking of meaning not as a property of the theory but as the field or context within which richer and more complex connections between sentences and things are established. Understood in this way, Cartwright's rejection of fundamentalism is Davidsonian in spirit: the "meaning" of a theory is its place within a more extensive pattern of interaction with other things.

What about those situations for which no plausible or accurate models are (yet) available? Cartwright argues that mechanics, for example, has no clear application to motions for which no force functions can be assigned. In the case of a thousand dollar bill blown about by the wind, she insists: "When we have a good-fitting molecular model for the wind, and we have in our theory systematic rules that assign force functions to the models, and the force functions assigned predict exactly the right motions, then we will have good scientific reason to maintain that the wind operates via a force. Otherwise the assumption is just another expression of fundamentalist faith."[41] Does this claim mean that most physical phenomena are not law-governed? Cartwright denies the claim that "no laws obtain where mechanics run out. Fluid dynamics [for example] may have loose overlaps and intertwinings with mechanics. But it is in no way a subdiscipline of basic physics; it is a discipline on its own. Its laws can direct the thousand dollar bill as well as can those of Newton and Lagrange."[42] When the laws of fluid dynamics and other disciplines lose clear application in turn, however, we may well encounter phenomena for which there is no more economical expression of what is happening than the phenomenon itself.[43]

---

together with the facts about the world in which it was articulated (the content of the theory would be those allowable models of the theory which offered reasonably accurate descriptions of the world in which the theory was actually put forward). This interpretation, however, is a far less interesting or even well-motivated theory than the one ascribed to Cartwright in the main text.

41. Cartwright 1994: 285.

42. Cartwright 1994: 284–85.

43. Cartwright would undoubtedly allow that there are still determining causes of such events, even without articulable general patterns or models through which those causes could usefully be classified. But even if Cartwright is correct about such singular causes, her point about laws can be extended to causes as well. One cannot assume that all events have causes except within the ranges where there is some reason to believe that those causes are identifiable.

The relevance of Cartwright's work to cultural studies begins to emerge once we ask what she means by a "model" of a theory or law. The term '*model*' has a very specific and constrained interpretation within what is sometimes called the *semantic* view of theories. But Cartwright clearly wants to use the term more loosely and widely than formalized interpretations of scientific theories could include.[44] Her own examples of models are mathematical, drawn largely from quantum mechanics and econometrics, but what matters for Cartwright is not their mathematical form but their use. Although she introduces them by analogy to logical empiricist "bridge principles" or "partial definitions," Cartwright sees models not as definitions of terms but as simulacra of things.[45] A simulacrum mimics features of the world which interest us in an object that we can manipulate in ways different from how we can manipulate the things simulated. In simulations as in the famous metaphor of the black box, only the links established at its beginning and its end matter. On Cartwright's account, simulacra establish connections between the formal structures of physical theory and the messy, concrete situations we encounter in the world. More precise, because we can establish one set of connections between the formal elements of a theory and its models and another set between the models and features of concrete situations that interest us, we are able to do things with both theories and things which could not be done without the models. Cartwright regards it as rare, although not impossible, for scientific theories to be mappable directly onto particular concrete situations without the intermediary of a simulacrum.[46] Mathematical simulacra are especially important in science only because of their economy of expression and well-developed repertoire of manipulative capabilities. They are also especially interesting because of the complicated practices (of which "measurement" describes only the tip of an iceberg) needed to connect their numerical or quantitative elements to features of the world.[47]

Once we understand the use of models as simulacra, a vastly wider range of things count as models in scientific practice. Models are new

44. Classical discussions of the semantic conception are provided by Sneed 1971 and van Fraassen 1980. Cartwright (1983: 159–62) briefly discusses how her own conception of models differs from the semantic conception.

45. Cartwright's principal discussion of simulacra is in 1983: 143–62.

46. Cartwright makes this claim in 1983: 160–61.

47. Cartwright's views about how mathematical theories are applicable to things in the world bears interesting affinities to Husserl's; for the comparison, see Rouse 1987a. Cartwright's account also meshes well with Latour's (1987: chap. 6) discussions of the work that must be done to establish and maintain connections between theoretical structures and other things.

objects that allow much denser interconnections between theoretical vocabularies and other objects and events in the world.[48] Although Cartwright highlights mathematical models, verbal, pictorial, and even physical models can play the same role. In the latter case, for example, recall the importance at one time of using wire and sheet metal contraptions to model the known constraints on the structural configuration of large molecules. Today, heterogeneous structural constraints are more likely to be combined and manipulated in a computer graphic model than in a literal physical structure, but the role of the model has not fundamentally changed. Of course, maps, diagrams, tables, photographs, and drawings have also been important for establishing similar connections between scientific vocabularies and objects in the world (think of anatomical drawings, geological survey maps, the periodic table, astronomical photographs, or Feynman diagrams of particle interactions).

What may initially seem to be a different kind of model is regularly constructed in scientific laboratories. Ian Hacking has emphasized the importance of laboratories in creating effects or "phenomena": discernable and noteworthy events or situations in the world. Some phenomena occur without deliberate intervention (rainbows, diurnal movements of the fixed stars, locust cycles), but many more are carefully and deliberately constructed in highly artificial settings. In *Knowledge and Power,* I characterized these settings as "microworlds": arenas in which things can undergo controlled interactions that are both intensively monitored and isolated from unwanted influences and contaminations.[49] The phenomena that manifest themselves in successfully constructed microworlds are also used as models intervening between verbal constructions and other things. One set of correlations between the verbal construction and the laboratory phenomenon and another set between the phenomenon and a variety of events outside the setting of the microworld enables a rich and powerful, albeit indirect, correlation between the unregulated events and the verbal practices. Even these dense and indirect correlations, however, can serve as truth conditions in Davidsonian interpretations of what people say and do.

An example may illustrate the interplay of various kinds of models

48. Hacking (1992) has provided a useful taxonomy of some of the heterogeneous elements that help attach theories more firmly to other things in the world. My point in extending the term 'model' is not to conflate what he distinguishes but to highlight the significance of his account for undermining representationalist accounts of how scientific theory relates to the world.

49. Rouse 1987b: chap. 4, 7.

in establishing truth conditions for sentences and thereby integrating those sentences with other things people do and say. Biologists now employ a rich terminology to articulate structural features of living cells: nuclei, ribosomes, mitochondria, membranes, Golgi bodies, and so forth. The application of these terms was regularized and is now learned through the use of multiple "models." Schematic diagrams depict these components in structural relationships. Those diagrams are connected to cells through various laboratory manipulations: discrete layers in ultracentrifugated cells and cell extracts; electron photomicrographs of appropriately stained and microtomed cell slices; and a multiplicity of other experimental constructions that isolate and display energy cycles and other exchanges with a cell's environment, genetic transcription, and a host of other biochemical processes within cells. With these models available, it is perfectly straightforward to learn to understand (and to utter understandably) sentences employing terms like "mitochondria" whose truth conditions encompass events taking place in unexamined cells outside the laboratory setting. Without these various diagrammatic, tabular, photographic, and material models, however, those sentences could not be interpreted (and therefore could not be said) to be "about" anything.

It should not be hard to see that being a model is not determined simply by the model's own properties or even by its relations to the specific things or situations it models. By themselves, models of all kinds (even verbal and mathematical models) are just more things in the world, with a multiplicity of relations to other things. What makes them models, and thus models *of* something, is their being taken up in practices, ongoing patterns of use for which norms of correctness and standards of success are applied with some regularity. The relevant conception of practices was developed in Chapter 5; here I want only to emphasize that I find no significant difference between worldly phenomena and verbal, pictorial, or mathematical models in their dependence on practices for their significance. Some phenomena, for example, are just curiosities. However noteworthy their occurrence, they are not informative about anything else. They only *become* informative when they are taken up in patterns of practice (with something at stake in the practices) which institute sufficiently robust, dense, and regular relations to other things.[50]

50. Mark Okrent alerted me to an important and relevant distinction that Dennett (1991a: 379–80) has made between two kinds of dependence of properties or relations on practices and understanding. Some properties are dependent only on the capacities of beings that understand (a vista is "lovely" if an appropriately constituted being would recognize its loveliness under the right circumstances, even if no such being ever actually occupies those

In many respects, it would be perfectly appropriate to substitute the word 'representation' for 'model' in the above discussion. If a representation is a thing that can stand in for another (and be recognizable as standing in for it) in a context in which it does not or cannot present itself (often to accomplish something that the things represented could not accomplish otherwise), then Cartwright's theoretical models and the illustrations, physical simulations, and phenomena that I assimilate to the concept of models are representations.[51] But such use is easily misunderstood, because the word 'representation' has accrued too many connotations from the notion of mental representation or linguistic representation that a Davidsonian semantics displaces. The objectionable connotations are at least threefold. First is the presumption that one thing represents another by belonging to a system of representation, for example, a language. When Davidson said "there is no such thing as a language,"[52] he meant that understanding is an open and dynamic engagement with the things around us, not containable within a closed system of rules or conventional relationships. Understanding can be partially modeled by truth theories for a language, but it always surpasses any systematic structure in ways that continually require a "passing theory" to accommodate. Second, the notion of mental representation fosters the illusion that representation is an objective relation between a representational content and what it represents instead of a pattern of use which only exists through its continuing reenactment. Third, because the physical "embodiment" of a representational token seems to involve a merely accidental association between a physical thing and its sense or content, the further illusion arises of an underlying pure expression or "voice" that confers meaning on the otherwise meaningless token.[53] Ultimately, these three points belong together: representation is at best a matter of one thing indicating another through its place in an open and ongoing pattern of practices, because there is no such thing as a language, a self-sufficient system of representation.

For the purposes of Davidsonian semantics, then, models are

---

circumstances or makes the recognition). Other properties depend on the actual exercise of capacities to understand (a person is a "suspect" only if someone actually suspects her). Being a model and thereby being informative about some other set of circumstances seems to me to be a property more like "suspect" than "lovely."

51. Latour (1987) has made extensive use of 'representation' in this way, partly to emphasize a connection between the semantic/epistemic and the political senses of the term.

52. Davidson 1986: 446.

53. This criticism of traditional concepts of representation has been most fully developed in Derrida 1973.

tropes. They do not express some meaning or content, but they do turn us in new directions by indicating differently. They encourage us to attend to patterns in the interpretation of sentences which differ from those that have been routinely employed in the past. Sentences such as 'His lecture was a real bomb' encourage attention to some of the generalizations we usually accept about bombs and about lectures, but not others. Likewise, treating hydrogen atoms as two-body systems for which only the Coulomb force is relevant encourages attention to some ordinary beliefs about hydrogen atoms and about the mathematical model, but not others.[54] Determining which routine beliefs should be brought to bear in a given case involves interpreting the force of the utterance, how it is supposed to be taken. 'Hydrogen atoms are two-body Coulomb systems' would be taken very differently if its readers thought it was intended to deny that electrons have spin. But the difference between ordinary predication and metaphors, models, or tropes is only one of degree. Every extension of a general term to a new case revises some of the generalizations that we were previously inclined to accept. Unless things have essential properties and properties have essential natures, sorting out the literal from the metaphorical is just deciding how to adjudicate between changing one's beliefs about the world and changing one's interpretation of a speaker.[55] The need for such adjudication can never be removed, because one cannot fix the rhetorical force of an utterance.[56]

This extended detour through Cartwright's account of models was intended to illuminate a possible connection between Davidsonian semantics and cultural studies of science. Cultural studies have commonly been directed toward texts, dramatic or cinematic productions, rituals, gestures, and other objects and practices whose significance can be located in the making and dissemination of meaning. If there were a sharp and principled distinction between literal predication and troping, the sciences might seem to be clear cases of literal reporting, whereas cultural studies would be appropriately directed toward the fictional and the tropic. But the point of denying that

54. Cartwright (1983: 136–39) discusses the relation between hydrogen atoms and relativistic models for the Coulomb potential between an electron and a proton.

55. Wheeler (1989: 116–39) develops such a Davidsonian blurring of any line between the literal and the metaphorical by emphasizing that all new predications involve changes in some previously accepted generalizations.

56. Wheeler (1989) concisely summarizes Davidson's argument to this effect: "A rhetorical force marker, a token that could be attached to an utterance to fix its rhetorical force, would, in virtue of being a mark at all, allow being used with different rhetorical force" (124).

distinction is not to assimilate one kind of practice or predication to the other: if science and "culture" do not have essential natures, then they are no more essentially the same than they are essentially different. The relativist move, which would demote scientific claims to fictions or elevate other cultural activities to "equal validity" with the sciences, commits this mistake: in its characteristic form, as we have seen, relativism or constructivism takes scientific statements and other cultural products to be alike in all being representations that are opaque to anything represented. Davidsonian semantics undercuts this move by asking how representations themselves could be any less opaque. Instead of representations intervening between us and the world, we have a world that includes sentences, models, and images and their practical and causal interactions with one another and with other things.

Cultural studies explore the heterogeneous interconnections among words, images, actions, and other events and things instead of reducing meaning to univocal representations of absent things. New ways of speaking, new kinds of image (cartographic, cinematic, computer graphic), and new manifestations of things illuminate one another and open new possibilities for understanding ourselves and the world. Such disclosures also foreclose or transfigure other previously familiar possibilities, and they realign the political solidarities and oppositions that enable and impede people's lives. Cultural studies are defined by the shift of attention to these transformations of the possible and the intelligible; their object is not some special domain of things (for example, meanings or representations) but the world as the setting for understanding and action. For Davidsonians, who insist that understanding discursive practices cannot be disentangled from understanding the things under discussion, no other way of locating cultural studies is plausible.

Nevertheless, one might object, the sciences deal with things in the world through such discursive practices as constructing theoretical models and manipulating phenomena in the laboratory; cultural studies deal with those scientific practices at one remove through other discursive practices of interpretation and cultural criticism. This objection is not mistaken as a characterization of the most immediate concerns and foci of attention in the sciences and in cultural studies of science. But scientists inevitably interpret and criticize their own practices in the course of doing science, while cultural studies must always deal with the circumstances in which scientific practices occur. The sciences and cultural studies of science have quite different histories, vocabularies, and styles of practice, but there remains substantial overlap in the objects, practices, and concerns that each must ultimately take into account. Cultural studies must be critically engaged with scientific practices and the world those practices belong to; the stances of reflective disengagement or deliberate estrangement from scientific practice which provide entree to the legitimation project cannot be sustained.

The rapprochement I propose between Davidsonian semantics and cultural studies of science may still face one formidable obstacle. Cultural studies often emphasize the importance of difference, power, ideology, and the proliferation of incommensurabilities, distortions, silencings, and other failures of understanding and communication. Davidsonian views, by contrast, initially seem to present understanding and communication as idyllic: everyone believes mostly truths and is mostly rational, while differences in language or conceptual scheme cannot suffice to permit massive failure of communication. But Davidsonians do not claim that speakers in fact agree in most of their beliefs or their norms of rationality. They present such agreement as an unavoidable presupposition of any interpretation of agents (including oneself). Such presuppositions are what enable differences to become manifest and understandable; without them, there can be neither agreement nor disagreement. Thus Davidsonians might contribute to cultural studies by providing the semantic resources for a richer account of difference and communicative breakdown.

Several strategies are available within a broadly Davidsonian approach for understanding some of the lacunae and obstacles within scientific understanding and communication (including the historical interpretation of earlier scientific work). First, one can recognize that understanding utterances cannot be disconnected from understanding their connections to other words and things. Without access to the models and images that often mediate these connections, the interpretation of verbal performances will likely be open to greater indeterminacy. The inability to reconstruct the instruments, procedures, and skills that create phenomena or to coordinate texts with images or diagrams imposes fewer constraints on truth-theoretical interpretations of what is being said. Davidsonians would similarly understand Hacking's suggestion that some sentences cannot even count as true-or-false in a particular setting. Sentences can fail to be true-or-false because of the inaccessibility of the "style of reasoning" that would connect those sentences to others in the kinds of patterns which would enable holistic, truth-theoretical interpretation.[57] Without the

57. Hacking 1982.

availability of patterns of statistical reasoning, for example, or of the recognized interconnections of resemblances and signatures in Renaissance natural philosophy, too much latitude exists for divergent interpretation of the sentences that characteristically fit into such patterns and interconnections and hence no clear basis for a determinate interpretation.[58] In these cases, lack of understanding follows from an insufficient evidence base.

Davidsonians can also recognize what we might call a "dynamic incommensurability" that follows directly from the lack of any magic language that would anchor meanings amid changing patterns of what is said when. Here the problem is not what has often been identified with incommensurability: the supposed untranslatability of utterances from one language or theory into another. Rather, although utterances can be interpreted only as part of a larger pattern of utterances, the utterances themselves carry no marker that could specify what pattern that is—what language is being spoken.[59] Thus gradual or localized changes in the patterns of linguistic practice will often not be immediately recognized. Ramberg has aptly spelled out the implications of this point: "Incommensurability is not a relation between [linguistic] structures, but a symptom of structural change. Such structural change can take the form of disintegration of a practice or paradigm, . . . or it can take the form of a merger of separate traditions. . . . Incommensurability *in discourse* can only begin to occur once we *think* we have begun to agree on linguistic conventions, but in actuality remain confused as to what language we are using."[60] Ramberg's claim is symptomatic of the overall Davidsonian approach to understanding and what makes it attractive for the purposes of cultural studies. Davidsonians see language as a dynamic process of making (and responding to) connections between sentences and other things. To adopt a useful metaphor from Ramberg, these connections have varying degrees of temporal, geographical, and social viscosity, variations that account for some of the resistance to perfectly fluid interpretation.

58. Styles or patterns of reasoning do not just cause indeterminacies in interpreting utterances that belong to those styles in a home language from which they are absent. The indeterminacy also runs to some extent in the other direction; after the introduction of new practices and norms of statistical reasoning, for example, earlier claims about probabilities might be made sense of in alternative ways. The problem is not that the previous statements once had a meaning that has now been lost. Nor is it just a problem of choosing the best translation for those sentences. The point is that utterances make sense only in relation to circumstances, and the circumstances in which they are interpreted have changed.

59. Recall that a "language" for a Davidsonian is nothing more than a set of actual and possible utterances for which there is a coherent truth-theoretical interpretation.

60. Ramberg 1989: 131–32 (first emphasis mine).

A final and crucial Davidsonian strategy for making sense of differences in understanding and blockages in communication is to recognize that language use is always embedded in larger patterns of activity. For Davidsonians, understanding utterances requires the interpretation of these more extensive practices. Utterances alone normally provide a limited basis for truth-theoretical interpretation of speakers; taking account of a broader range of actions affords more extensive grounds for attributing beliefs and desires. Of course, all actions, including utterances, are situated amid the anticipated and actual responses of other agents. Interpreting an agent thus minimally requires taking account of the agent's interpretation of others. But here we also encounter the dynamics of social power: understanding an agent requires understanding how various agents' actions align with one another and their environment to maintain asymmetries of power and how the interpreted agent can act to reinforce or avoid the consequences of those asymmetries.[61] We have already seen that other speakers can assimilate unfamiliar patterns of speech to more familiar and perhaps less threatening uses. On the other hand, one way to resist social power is to communicate and act in ways that avoid making oneself fully transparent to interpretation. Irony, ambiguity, tropes, and silence all belong to speakers' repertoires for such indirect communication.[62] For Davidsonians, the only constraint on either assimilation or the rhetorical subtleties of resistance is that they must be intelligible to an interpreter (perhaps one more imaginative or differently situated than those to be deceived) on the basis of the publicly available evidence. There can be no deeper basis for saying that an avowal really was ironic or a silence meaningful than the coherence of the overall interpretation of what was said and done.

Davidsonian strategies thus provide considerable resources for understanding disagreement, misunderstanding, and incommensurability. These strategies remain subordinate, however, to the fundamental Davidsonian insistence that understanding meaning and cultural practices cannot be disentangled from understanding the circumstances in which those practices take place, and vice versa. I have argued that this central theme of Davidsonian semantics offers two fundamental advantages for contemporary science studies. First and foremost, it undercuts the plausibility of representationalist

61. Wartenburg (1990) offers a useful discussion of social power relations as dynamic. See also Chapter 7, above.

62. For a persuasive account of the construction of scientific incommensurabilities as a deliberate response to asymmetries in social power (which were also epistemically consequential asymmetries), see Biagioli 1993: chap. 4.

epistemologies and the legitimation project, thereby making the issues in dispute among realists, rationalists, and social constructivists seem less interesting and less relevant to understanding scientific practices and their achievements. Second, it supports a constructive program for critical engagement with scientific practices. Cultural studies of science do not attempt to survey scientific practice as a whole and pronounce on its aim and legitimation. Cultural studies instead address particular discursive practices and the specific interactions with other practices and things through which those practices become significant. The importance of Davidsonian semantics for cultural studies of science is to encourage the critical examination of discursive practices in the sciences as a dynamic and ongoing engagement with the world rather than as enclosed and determinate systems of representation.

[9]

What Are Cultural Studies of Science?

The social constructivist tradition has dramatically transformed interdisciplinary studies of the sciences over the past two decades. The postpositivist interdisciplinary formulation of "history and philosophy of science" has been fundamentally challenged by the sociological perspectives offered by the Edinburgh Strong Programme, the Bath constructivist-relativist approach, applications of discourse analysis to science, and ethnographic laboratory studies. Many features of scientific work which have been highlighted by these sociological traditions have become indispensable considerations for subsequent interpretations of scientific work. Social constructivist studies have brought renewed attention to the epistemic importance of laboratory practices and equipment, to the omnipresence of conflict and negotiation in shaping the outcome of scientific work, to the formation and dissolution of disciplinary boundaries, and to the permeability in practice of any demarcation of what is "internal" to scientific work. Constructivist studies have also effectively highlighted the sheer difficulty of scientific work: getting equipment and experiments to work reliably, replicating their results, and achieving recognition of their success and significance.

Despite the significance of social constructivism, however, much subsequent work in science studies does not easily fit within the legitimation project that encompasses the disagreements between social constructivists and the proponents of internalist history and philosophy of science. Among the central issues between social constructivists and internalists were the relative importance of social and rational (or external and internal) "factors" in explaining the content of scientific knowledge, the relations between empirical descriptions and epistemic evaluations of the methods and achievements of scientific research, and the coherence of either realist or relativist/constructivist accounts of how scientific knowledge is related to the world. Work in a variety of science studies disciplines has increasingly challenged the very terms of these debates. Concerns have been raised about the goal of explaining scientific knowledge, the presumed explanandum of the "content" of knowledge, the supposed opposition between descriptive and normative approaches, and the intelligibility of the question that realist or constructivist interpretations of knowledge are supposed to answer.

In this chapter, I shall articulate and illustrate some important issues that mark the movement beyond the terms of the disputes between internalists and social constructivists. I introduce the phrase "cultural studies of science" to refer to this quite heterogeneous body of scholarship in history, philosophy, sociology, anthropology, feminist theory, and literary criticism. In using such a phrase, we must keep in mind that it cuts across some very important theoretical, methodological, and political differences and that some significant scholarly work takes place across the very boundaries I articulate between cultural studies and the social constructivist tradition. My aim is not to reify cultural studies but to highlight some important issues that might reshape the terms of interdisciplinary science studies.

So what are *cultural studies of science?* I use the term broadly to include various investigations of the practices through which scientific understanding is articulated and maintained in specific cultural contexts and translated and extended into new contexts. '*Culture*' is deliberately chosen both for its heterogeneity (it can include "material culture" as well as social practices, linguistic traditions, or the constitution of identities, communities, and solidarities) and for its connotations of structures or fields of meaning. Several historical vignettes may help situate the differences between cultural studies as I conceive them and the sociological and philosophical traditions to which they are responding. I should emphasize that these sketches constitute only some possibly revealing fragments of a history of cultural studies. In this context, I shall then discuss more systematically what I take to be the most important theoretical issues that demarcate cultural studies of science as a significant and distinctive field of inquiry.

My first historical note fittingly recognizes the indebtedness of cultural studies of science to the social constructivist tradition. Cultural studies follow the lead of the Strong Programme and its sociological successors in refusing to require distinctive methods or categories to understand scientific knowledge as opposed to other cultural formations. Karl Mannheim's earlier sociology of knowledge notoriously exempted the natural sciences and mathematics from its purview.[1] Similarly, the Mertonian tradition, which still largely dominates American sociology of science, did address the natural sciences but insisted that its investigation of scientific institutions and norms largely took for granted the content of successful scientific work.[2] Mertonians have been concerned over how that work could be embodied institutionally and culturally and how deviations from its established norms and methods might be appropriately explained. Much of philosophy of science (and some historical work) have likewise been constituted by distinctions between (on one hand) the imagination, reasoning, and evidence "internal" to the establishment of scientific knowledge and (on the other hand) the biographical and social factors that at least ideally might be excluded from epistemological reflection. Cultural studies may nevertheless go further than some social constructivists in refusing to make knowledge, or the "content" of knowledge, an important *focus* of their inquiries: a deflationary approach to knowledge thereby departs from any attempt to characterize knowledge more generally in terms of consensus, representation, or rule-governed forms of life.

By contrast, cultural studies of science take as their object of investigation the traffic between scientific inquiry and those cultural practices and formations that philosophers of science have often regarded as "external" to knowledge. The sciences are taken to be cultural formations that must be understood through a detailed examination of the resources on which their articulation draws, the situations to which they respond, and the ways they transform those situations and have an impact on others. As I shall argue, cultural studies do not try to replace internalist accounts of knowledge by relying on a privileged alternative explanatory framework (for example, social factors), but neither do they grant epistemic autonomy to what is currently accepted as scientific work.

A second, more historically specific vignette may help locate some interesting differences between social constructivism and cultural

1. Mannheim 1952.
2. See, for example, Merton 1973.

studies of science. The culture and politics of scientific knowledge became a focal point of state politics in the United States and Great Britain during and after the Second World War, as the state became more actively involved in the support and direction of scientific research. The issue broadly concerned how best to organize, support, and direct scientific inquiry in a democratic political culture. In Great Britain, crystallographer J. D. Bernal argued for the deliberate political management of science for socially beneficial ends.[3] Bernal was a committed socialist who argued that a capitalist society was incapable of developing or utilizing scientific knowledge effectively or humanely. He emphasized that scientific inquiry was a social product of human labor, which required considerable resources, and that it promised great benefits but could also create new resources for oppression. What was needed was a social transformation in which a humane science could flourish, one that he also saw the aims of science itself call for implicitly: "Science implies a unified and coordinated and, above all, conscious control of the whole of social life."[4]

"Bernalism" was prominently opposed by the physical chemist Michael Polanyi.[5] Polanyi's epistemology emphasized the importance of practical skills and nonverbal communication in what he called the "personal knowledge" that shapes scientific work. But his position had important and conservative political consequences. Science could not be deliberately directed to social ends without undermining its epistemic success; furthermore, because the basis of scientific knowledge was inarticulable, no one could understand how best to advance science who was not a practicing scientist. Polanyi saw no alternative to unrestrained freedom of scientific inquiry, and administrative control of scientific resources by a scientific elite.

The social constructivist tradition has taken an ambivalent stance toward the Bernal/Polanyi debate. Constructivists have adopted a Bernalist interpretive stance toward scientific activity, emphasizing that research is a process of social production and certification which must be understood in terms of social categories. The descriptions of scientific activity which they have developed, however, are deeply indebted to Polanyi. Polanyi's account of scientific knowledge as locally situated, tacit know-how directly influenced both relativist and ethnographic studies of scientific laboratories (for example, those by Harry Collins, Trevor Pinch, Bruno Latour and Steve Woolgar, Karin

3. Bernal 1967, 1954; for an informative discussion of Bernal's views, see Werskey 1978.
4. Bernal 1967: 409.
5. Polanyi 1958.

Knorr-Cetina) which have been an important component of the constructivist tradition. Furthermore, despite their occasional rhetoric of antiscientism, the constructivist tradition has predominantly shared Polanyi's antinormative stance, which forecloses the possibility of criticizing scientific practices and beliefs.[6] Constructivists initially seem to preclude criticism of scientific practices on different grounds than did Polanyi; they espouse a far-reaching epistemic relativism instead of an elitist defense of the unquestionable authority of scientific communities. Yet, in practice, these two positions converge in their defense of community authority. Thus, constructivists Collins and Steven Yearley offer this Polanyiesque objection to Michel Callon's account of the fate of a French research project on scallop cultivation: "There is only one way we know of measuring the complicity of scallops, and that is by appropriate scientific research. If we are really to enter scallop behavior into our explanatory equations, then Callon must demonstrate his scientific credentials."[7]

Where social constructivists therefore find themselves drawn to both sides of the Polanyi/Bernal debate, proponents of cultural studies will typically be attracted to neither. The poststructuralist theoretical influence on much of cultural studies of science is not congenial to the Marxist humanism that animated Bernal: Bernal's presumption of a common human interest and a shared project of liberation through the social appropriation of production is at odds with the sensitivity of cultural studies to differences and contested meanings and identities. Yet Polanyi's vision of a self-managing scientific elite is still less attractive. Instead of sanctioning or relativizing scientific communities, cultural studies contest their boundaries and the authority established by marking and policing those boundaries.[8] A very different politics of knowledge must follow from this stance, neither Polanyi's scientific oligarchy nor constructivists' pluralism of epistemic communities.

Such an epistemic politics cannot allow the scientific community to speak authoritatively in a unified voice; nor can it colonize science in the name of a privileged vocabulary imposed on science from a standpoint of epistemic sovereignty.[9] My final historical vignette thus

6. For a useful discussion of the continuity between the social constructivist tradition and Polanyi's antinormative account of scientific research, see Fuller 1992c.

7. Collins and Yearley 1992a: 316.

8. For critical discussions of the epistemic significance of policing epistemic borders, see Traweek 1992 and Rouse 1991c.

9. The concept of epistemic sovereignty was introduced to extend Foucault's (1978, 1980) criticism of the ways in which the problematic of sovereignty frames political theory. The concept is articulated and critically applied to science studies in Rouse 1993, 1994.

appropriately emphasizes the indebtedness of cultural studies of science to the last half-century of political criticism of science from within the scientific community. Contemporary cultural studies of science owe much to the political ambivalence among physicists which led to the *Bulletin of the Atomic Scientists* as well as to the more widespread scientific opposition to militarized scientific research (especially during the Vietnam War); the formation of groups such as Science for the People and the Radical Science Journal Collective; the rise of a scientific environmentalism that included opposition to corporate and government domination of research on pesticides, low-level radiation, and so forth; up through the controversies over recombinant DNA research and then the Human Genome Project. The first wave of research on issues of science and gender, which emphasized the criticism of ideological treatments of gender in biology and psychology, was also largely the work of scientists, and their work was probably a precondition for the more far-reaching discussion of science and gender in cultural studies.[10] Cultural studies of science belong not only to the history of the academy and its disciplined historical, philosophical, and sociological interpretations of science but also to the historical studies of science, the culture of science itself, and political struggles over scientific knowledge.

In situating cultural studies of science in these ways, I have tried to emphasize their continuity with important aspects of the twentieth-century culture of science. But now the time has come to say something about their own distinctive contributions to understanding science. Of course, given that cultural studies of scientific knowledge are both diverse and contested, I find something artificial about attributing to them a common picture of scientific work. Yet there are significant common themes, however diversely developed, that mark important contrasts to other ways of understanding the sciences. I shall mention six such themes: antiessentialism about science, a nonexplanatory engagement with scientific practices, an emphasis on the locality and materiality of scientific practices and even more so on their cultural openness, subversion of rather than opposition to scientific realism or conceptions of science as "value-neutral," and a commitment to epistemic and political criticism from within the culture of science.

Cultural studies of science reject the existence of an essence of science or a single essential aim to which all genuinely scientific work

10. For example, Hubbard 1990, Fausto-Sterling 1985, Bleier 1984 and Birke 1986.

must aspire. In Richard Rorty's succinct formulation, "Natural science [is not] a natural kind."[11] The practices of scientific investigation, its products, and its norms are historically variant. They also vary considerably both across and within scientific disciplines. High-energy physics, low temperature physics, radio astronomy, synecology, molecular biology, taxonomy, paleontology, and meteorology are in many respects quite different epistemic practices, and this list does not encompass even more directly "applied" scientific fields. Scientific work is also culturally variant even within the same field; there are often, for example, important national differences in the style, direction, standards, and goals of scientific work.[12] This does not at all mean that different scientific cultures are self-enclosed or mutually uncomprehending or that individual scientists or groups cannot navigate their borders quite effectively. Nor does it mean that the epistemically interesting differences in scientific cultures neatly map onto national, linguistic, or other cultural boundaries.

For now I just want to emphasize that the variability within scientific practice involves many of its important features. It includes the scale, precision, technological sophistication, sensitivity, theoretical transparency, and theoretical independence of its instruments; the scale, location, mobility, and accessibility of its objects of inquiry; its social order (for example, the size of its effective research groups and their degree of heterogeneity in knowledge, skill, mutual understanding, status, and so on); its theoretical sophistication and the relations between theory and experimental or observational practice; its distance from specific "applications" of knowledge; the character and significance of its engagement with other cultural practices; the relative importance of description and explanation; and the institutional organization of its research and communication.

Insensitivity to the heterogeneity of the sciences is an important part of what cultural studies take to be wrong with global legitimations of the rationality of science or of its referential success and what they take to be equally wrong with those epistemic relativisms that place scientific communities (and their accepted results) on a par with others and with one another. Whether one argues that scientific inquiry as such is superior to other epistemic practices, is "no better than" others, or is somehow less adequate, the mistaken assumption once again is that scientific knowledge belongs to a single kind similar

11. Rorty 1991: 46–62

12. Recent studies of such differences include Traweek 1988 and Harwood 1993.

or distinguishable *in kind* in any interesting way from other kinds. Similar problems are manifest in any attempt to distinguish natural science from social or human science.[13]

The antiessentialism of cultural studies extends to my second theme. One of their most important differences from the social constructivist tradition is their opposition to an explanatory stance toward scientific knowledge (or its "content"). Social constructivism typically presents itself as an explanatory social science that can (potentially) account fully for the epistemic outcomes of scientific practices. In this case, the vocabulary of social interaction (interests, negotiations, and so on) is supposed to hold the key to an adequate understanding of scientific work. But as Nancy Cartwright has noted about physical explanation, "the aim [of an explanatory science] is to cover a wide variety of different phenomena with a small number of principles. The explanatory power of [a] theory comes from its ability to deploy a small number of well-understood [expressions] to cover a wide variety of cases. But this explanatory power has its price [which is] to constrain our abilities to represent situations realistically."[14] The need to account for the phenomena in terms of a theory's explanatory concepts suppresses differences among the phenomena being explained, whether those differences are susceptible to alternative explanatory frameworks or not. For example, a social explanation of the content of a scientific practice is not well situated to consider the variety of ways such a practice may be appropriated and used; cultural studies of science may well be concerned with the plasticity of what constructivist studies take as an unproblematic explanandum.

But two related difficulties with an explanatory stance are perhaps even more fundamental for cultural studies. First, cultural studies take exception to the ways in which an explanatory stance reifies the boundaries between the interpretation and what it interprets. This reification can take different forms. Latour and Woolgar, for example, adopt (at least rhetorically) the stance of the ethnographer as stranger,[15] whereas Collins and Yearley present themselves as disciplinary antagonists to the natural sciences: "We provide a prescription: stand on social things—be social realists—in order to explain natural things. The world is an agonistic field (to borrow a phrase from Latour); others will be standing on natural things to explain

13. Rouse 1987b: chap. 6.
14. Cartwright 1983: 139.
15. Latour and Woolgar 1986.

social things. . . . [SSK, then,] wants to use science to weaken natural science in its relation to social science."[16]

Cultural studies have instead been influenced by that tradition in postcolonial anthropology which is suspicious of attempts to impose categories unilaterally on the Other,[17] even when anthropology has been repatriated, science has been made into the Other, and the imperializing anthropologists present themselves as the "underdog" to the established cultural authority of the natural sciences.[18]

The second related problem with social explanations of scientific knowledge concerns the reification of the categories of the (social) explanans, which is self-consciously defended by Collins and Yearley in the passage I just quoted. Cultural studies focus on the articulation and significance of meanings and are reluctant to set the categories of social explanation outside of their purview. This reluctance increases wherever such explanations presume the unity of social identities or categories, which cultural studies must frequently deconstruct. Such an exception becomes more troubling given the widespread acknowledgment that the categories and practices of social explanation themselves belong to a scientific tradition.[19]

Such reifications of the categories of a social science have been critically discussed within the social constructivist tradition under the heading of "reflexivity." This issue has been most prominently associated with the work of Woolgar and of Malcolm Ashmore.[20] Woolgar and Ashmore regard the aspiration to social explanation of scientific knowledge as rhetorically incoherent: sociological accounts aim to

16. Collins and Yearley 1992b: 382–83.

17. For discussions of these issues, see Marcus and Fischer 1986, Clifford and Marcus 1986, and Rosaldo 1989.

18. Collins and Yearley (1992b) explicitly portray their own explanatory antagonism toward the cultural hegemony of the natural sciences as like "the underdog so familiar from romantic portrayals of science" (382).

19. For many social constructivists, sociology offers an alternative route to a fully naturalized scientific study of science itself. Whereas Quinean naturalists, for example, take epistemology or philosophy of science to be the scientific study of how scientific knowledge is constructed from sensory stimulations, the Edinburgh school finds the crucial issue for a scientific study of science in how diverse beliefs are generated from similar environmental surroundings. In both cases, however, the aim is to close the domain of science by bringing its own activities within its purview, and in both cases, the resulting knowledge stands or falls with the claims of scientific knowledge more generally. The critical resources of such naturalized epistemologies are limited to identifying and removing inconsistencies within the totality of science. Cultural studies take these critical resources to be inadequate. A naturalized science of science is also thereby committed to a conception of science as essentially aimed toward producing a consistent system of representational knowledge, a conception that cultural studies reject.

20. See esp. Woolgar 1988b,c and Ashmore 1989.

undermine the naîvete and apparent transparency of scientific representations, but they are no less naîve in their own representationalist rhetoric. In the previous chapter, I argued against the representationalist conception of language which frames Woolgar's presentation of reflexivity and against the general philosophical skepticism that his account suggests. Cultural studies, however, respond to reflexive criticism of the aspiration to social explanation in ways that can be further differentiated from Woolgar's and Ashmore's approaches. For Woolgar and Ashmore, taking reflexivity seriously suggests giving up the instrumental concern to improve representations *of* science in favor of inventive and playful "interrogations" of representation itself: "Reflexivity," Woolgar proclaims, "is the ethnographer of the text."[21] Cultural studies instead take reflexive questions as an invitation to consider their own complex epistemic and political relations to the cultural practices and significations they study.[22]

This distinction is not meant to exempt the rhetoric of science studies from reflexive criticism. Donna Haraway and Sharon Traweek, for example, also criticize common rhetorical strategies within science studies, but to different ends. Traweek notes that, like scientists, "almost all those writing the newer social studies of science and technology also account for everything and reject all other stories. Almost all these stories, whether about nature, scientists, or science, are narrative leviathans, producing and reproducing all-encompassing stories of cause and effect through the same rhetorical strategies."[23] Such stories work by "relentless, recursive mimesis; the story told is told by the same story."[24] In criticizing such rhetoric, Haraway and Traweek are not concerned to "interrogate" and defamiliarize representational practices generally. From their perspectives, Woolgar's stories make up yet another narrative leviathan, one about how all representations (including his own) are projections of "the Self." For Haraway and Traweek, by contrast, reflexivity discloses partiality and situatedness, not self-enclosure. It exposes the illusion that representation is autonomous and self-projecting; we can never encounter or understand ourselves except through our interactions

21. Woolgar 1988b.

22. Haraway 1989 and Traweek 1992 both illustrate how reflexive attention to the construction of one's own text can engage the political significance of doing cultural studies of science and not just undertake the futile rhetorical task of representing the supposed "ideology of representation" (Woolgar 1988c: chap. 7).

23. Traweek 1992: 430.

24. Haraway forthcoming: ms. 14.

with others in partially shared surroundings. If rhetoric is always situated, then reflexive concern for one's own authorship cannot remain internal to the text. The textual self-presentation of the author is subject to reflexive criticism only as part of a larger concern for writing and speaking as forms of action. What do these writings and sayings do? To whom and about whom are they expressed? In what ways do they allow for and acknowledge, or foreclose and not hear, the responses of those to, about, or past whom they speak? Above all, to whom are they accountable? Critical reflection on knowledge claims is thus always as much moral and political as epistemological, and Haraway's and Traweek's calls for reflexivity aim to reconfigure the politics of science and science studies. Haraway notes that while "the natural sciences are legitimately subject to . . . cultural and political evaluation 'internally,' not just 'externally,' the evaluation is also implicated, bound, full of interests and stakes, part of the field of practices that make meanings for real people accounting for situated lives."[25] Reflexive attention to one's own practices of speaking and writing is thus integral to a political engagement with science which would be appropriately modest and self-critical.[26]

Haraway's and Traweek's emphasis on reflexively situated inquiry points toward a third distinctive feature of cultural studies, an insistence on the local, material, and discursive character of scientific practice.[27] Scientific knowledge is often discussed as if it were a body of free-floating ideas detachable from the material and instrumental practices through which they were established and connected to things. Cultural studies (along with other recent studies of experimental practice) instead emphasize the importance of specific complexes of instruments and specialized materials, as well as the skills and techniques needed to utilize them, in shaping the sense and significance of knowledge. They also emphasize the particularity of networks of scientific communication and exchange which shape both what needs to be said and what vocabulary and technical resources can be appropriately utilized. For example, cultural studies stress the

25. Haraway 1989: 13.

26. Haraway's discussion of the scientific career of primatologist Alison Jolley (1981: chap. 10) provides an illuminating example of how such reflexive modesty might be realized in one very particular setting. Jolley's career and its scientific and cultural setting are unusual in ways that would strongly discourage taking her work as a model for politically engaged scientific practice, but it nevertheless illustrates Haraway's conception of a reflexive rhetoric and politics.

27. For a detailed discussion of the locality and materiality of scientific knowledge, see Rouse 1987b: chap. 4, and Rouse 1991c.

ways in which disciplines can be created or transformed as much by new instruments and objects as by new concepts or theories (although we should beware of distinguishing these categories too sharply, as if instruments and objects were somehow prediscursive). The transformation of classical cytology into modern cell biology was focused more by uses of the ultracentrifuge and the electron microscope than by any particular theoretical innovations, but it thereby changed what counted as a scientifically interesting question about cells and an adequate answer to this question.[28] Peter Galison has argued as well that some basic concepts of particle physics were altered by the use of counters in the 1930s; in practice they transformed "electron," for example, from an aggregate to an enumerable concept (without instantiating distinct individuals).[29]

Instruments belong ineliminably to local contexts within which there are the facilities, skills, and discursive practices that enable them to operate significantly. Philosophers in the 1960s and 1970s thought that the influence of instruments on scientific knowledge could be captured in terms of the theory-ladenness of observation, which presumed that the crucial aspects of the instrument's functioning were theoretically understood. Almost invariably this is not the case, as sources of error and noise are regularly circumvented by practical engineering that does not require full theoretical comprehension.[30] The locality of knowledge is also suggested by the importance of the exchange of actual materials to be used or investigated (particular cell cultures, plasmids, superconducting ceramics, and so on); they are not readily reproducible from a description. Some scientists and philosophers may balk at this emphasis on the irreducible locality of scientific knowledge, but they should be clear about what they are thereby doing: they are excluding from scientific knowledge most of what experimentalists, instrumentalists, and even phenomenologists within the sciences distinctively know and do.

Cultural studies' emphasis on the locality and materiality of scientific practices must nevertheless be distinguished from the suggestion that such practices exhibit knowledge that is "tacit" (as Polanyi argued) or mute (as is perhaps implied by some studies of experimental

28. For detailed discussions, see Rheinberger 1992c, 1994, and Bechtel 1993. The differences between these discussions are themselves salutary for the concerns of the present argument, for Rheinberger's account exemplifies the kind of interpretation of the articulation of meaning which I attribute to cultural studies, whereas Bechtel attempts a mediation between social constructivism and more-traditional history and philosophy of science.

29. Galison 1987: chaps. 2–3.

30. Hacking 1983; Traweek 1988: chap. 2.

practice, which may seem to suggest a materialist explanation of scientific knowledge as opposed to its cultural interpretation). In either case, material practice would be rendered *inarticulable* and hence inaccessible to the interpretive practices of cultural studies. The localization of scientific practices and capabilities should also not be taken to exclude the adaptation and standardization of practices to extend them into new local settings and to establish and maintain large-scale continuities across those settings.[31] Latour's account of the extension and maintenance of networks and their connections to centers of calculation[32] and my earlier discussion of the dissemination of knowledge and power through contested alignments are quite consistent with the locality of scientific practice. The claim that scientific practices and knowledges are local is opposed to conceptions of the effortless and immaterial universality of scientific reasoning and knowledge, not to the specificity and materiality of global interconnections (networks, alignments, relations) which might be extended anywhere (with sufficient social and material support) but which can never hold everywhere at once.[33]

My fourth theme from cultural studies, what I have been calling the openness of scientific practices, conflicts with a widespread sense of scientific communities as relatively self-enclosed, homogeneous, and unengaged with other social groups or cultural practices. Even such an influential and informative precursor to cultural studies of science as Thomas Kuhn's *Structure of Scientific Revolutions* stresses the intellectual and normative autonomy and uniformity of scientific communities. The social constructivist tradition has often followed Kuhn in this respect, focusing on either the social interests or the social interactions that constitute the shared beliefs, values, and concerns of scientific communities.[34] But cultural studies of science display a constant traffic across the boundaries that allegedly divide

31. Rouse 1987b: chaps. 4 and 7.

32. Latour 1987: chap. 6; for the purposes of cultural studies, Latour's account of networks, centers, and the metrologies and other policing practices that sustain them may be more attractive than his semiotic interpretation of those practices and his rhetorical construction of technoscience and science studies alike as militarized trials of strength. For critical discussion of these latter aspects of Latour's work, see Haraway 1994, forthcoming, and Lenoir 1994.

33. Fraser and Nicholson (1988) usefully discuss the political significance of rejecting universality without abandoning the analysis of particular large-scale historical and geographical constructions.

34. Kuhn 1970; prominent examples of social constructivists who emphasize the role of relatively enclosed scientific communities or forms of life include Collins 1992 (especially his notion of a "core set") and Bloor 1983. Knorr-Cetina 1981, with its emphasis on "trans-scientific fields," was remarkable for its early divergence from the focus on scientific communities.

scientific communities (and their language and norms) from the rest of the culture. Latour has provocatively expressed this sense of the openness of scientific work in saying that scientific work itself effectively destabilizes any distinctions between what is inside and outside of science, or between what is scientific and what is social.[35]

It is important to recognize that the traffic across the boundaries erected between science and society is always two-way. For now, I will talk about the ways in which scientific work continually draws on and is influenced by the culture "outside." The traffic in this direction involves, among other things, scientists seeking and acquiring material and financial resources, recruits, meaningful or significant questions and problems to investigate, a vocabulary and the metaphors and analogies it incorporates, allies, and much more. I want to present multiple examples to illustrate my point, to make plausible the range and depth of the claim that cultural studies make and their justification for it.

My initial case is taken from Robert Marc Friedman.[36] Friedman has shown how important *theoretical* features of Vilhelm Bjerknes's evolving studies of atmospheric geophysics were shaped by specific relationships cultivated with military and civil aviation, fisheries, and agriculture. Bjerknes's group replaced the prevailing statistical/climatological approach to meteorology with a three-dimensional modeling of atmospheric dynamics. His models emphasized the formation and movement of atmospheric discontinuities (or "fronts"). But this very conception initially depended on both the needs of and the resources providable by aviation and shipping. Aviation needed much finer-grained and differently conceptualized atmospheric analyses than prevailing meteorological theory could discriminate. Yet airplanes and airships were necessary for acquiring data enabling a three-dimensional atmospheric geophysics that could reveal rapidly moving and sharply delineated discontinuities. These relationships were indispensable for the successful struggle to impose a common instrumentarium and metrology throughout Europe and North America, one marked in physically rather than phenomenologically significant units and temporally synchronized rather than timed for local convenience. Most previously practicing meteorologists did not even comprehend the new units of measurement; yet, without these changes, there could be no knowledge of relevant atmospheric features.

35. Latour 1983, 1987.
36. Friedman 1989.

High-energy physics (HEP) may seem more remote from particular social interests or cultural practices than does meteorology. But cultural and political engagement can make considerable difference in what kind of knowledge can be produced. As Traweek has pointed out, the principal determinant of an HEP group's work is its detector.[37] All accelerator research groups take pulses of particles from the same beam, but what knowledge they produce depends on the detector they put in its path. In the United States, detectors are short-lived and continually tinkered to keep them at the very edge of the state of the art without encountering irreducible noise in one's data or expense and time in one's work. Experimental physicists build detectors themselves (and rework them), both to minimize noise and to achieve the precise data response desired. In Japan, this approach is not possible. Funding for HEP is tied to Japan's corporations; physicists specify only general design criteria for a detector, which is then built by industrial firms and cannot be altered on site. As Traweek notes, such highly expensive machines with the most sophisticated components must then be used for a long time. Whereas in the United States a physicist will typically work with several generations of detectors, in Japan a detector will survive through several cohorts of physicists who spend their careers with one machine. These differences strongly affect the kinds of questions one can ask and the characteristics of good results.

My third example comes from historian Donna Haraway. She has documented a sharp transformation in the 1940s and 1950s in the metaphors that organized research and its interpretation in several fields of biology, notably evolutionary theory, genetics, developmental biology, and immunology. Haraway described the change as "a transformation from a discourse on physiological organisms, ordered by the hierarchical sexual division of labor and the principle of homeostasis, to a discourse on cybernetic technological systems, ordered by communications engineering principles."[38] Haraway's argument connects both the theoretical and economic resources for these transformations of core fields of biological science to war-related developments in operations research and labor management, and their intellectual plausibility in part to contemporary transformations in the economy and in cultural images of language and self. Such metaphorical structures in science are tremendously important epistemically, especially because of the ways they shape the develop-

37. My discussion of this example is drawn from Traweek 1988.
38. Haraway 1981–82: 245.

ment of subsequent research. They help frame the interesting questions and what would be intelligible as an answer to those questions.

The intertwining of scientific knowledge with cultural constructions of sex and gender should be specially emphasized, for it has been very influential in the formation of cultural studies of science. Some engagements of science and gender should by now be unsurprising (although they have certainly not been uncontested). Could research into endocrinological influences on sex differences in behavior or ability or evolutionary explanations of gender difference be expected to escape the effects of cultural constructions of gender? Similarly, when one recognizes the epistemic importance and cultural complexity of researchers' credibility, it would be astonishing if gender were not significant there. I thus choose two more indirect examples to accentuate the theme of the openness of scientific work.

The first comes from Evelyn Fox Keller, whose historical inquiries have concerned the cultural formation of molecular biology in its peculiarly central place within the biological sciences today. From H. J. Muller's ecstatic analogies between his x-ray-induced genetic mutations and Rutherford's bombardment of atomic nuclei with alpha particles ("Mutation and transmutation—the two keystones of our rainbow bridges to power"[39]) to molecular biologists' frequent identification of DNA molecules with "the secret of life" and "the displacement of flesh-and-blood reference that is [thereby] symbolically effected,"[40] Keller argues that the representation of the significance of molecular biology has been powerfully gendered. She interestingly connects the ways scientists have attempted to legitimate the biological centrality of this work to powerful cultural narratives of male birthing and second birthing. What is at issue here is not the specific role that DNA molecules play in heredity but the gendered significance of specific research programs in biology in relation to other elements of biological (and physical) science.

A very different sort of example is displayed in a recent discussion by Haraway of the content of *Science*.[41] The contents of this official journal of the American Association for the Advancement of Science are usually understood to be its scientific articles and its letters, news, and commentary. But almost a quarter of the journal's actual pages by my count are typically devoted to advertisements. This fact alone

39. Quoted in Keller 1990: 397.
40. Keller 1992: 52.
41. Haraway 1992b.

suggests the economic significance of scientific practices and equipment. But what Haraway has done is to study the imagery developed and exploited in the advertisements to striking effect. From the rabbit at the computer keyboard staring at its graphically constructed image on the screen ("A few words about reproduction from an acknowledged leader in the field") to the male scientist bottle-feeding a monkey in the lab at midnight and to DuPont's genetically engineered laboratory mouse with active oncogenes ("OncoMouse™"), the humor and imagery in the advertisements play subtle and not-so-subtle variations on cultural narratives of gender and birthing, origins and salvation, purity and pollution, nature and culture. These advertisements raise complicated issues about their intended audience and the significance of the imagery they embody. But they remind us that scientific understanding encompasses much more than what appears in the carefully dry prose of the canonical journal report.

This sense of the openness of scientific practices and the inappropriateness of any principled boundaries between what is internal or external to science brings us back to the close linkages between knowledge and power. The discussion in Chapter 7 showed that knowledge and power cannot be reduced to the same thing, because neither should be understood as a kind of thing at all. Talking about knowledge and about power are ways of understanding the interrelations among practices and the things disclosed within them. 'Power' and its associated concepts (coercion, domination, authority, empowerment, and so forth) provide a way to understand how some agents and instrumentalities act in concert within a particular setting to transform others' possibilities for acting. The use of 'knowledge' and related concepts enable interpretation and assessment of how agents and their surroundings function together to disclose or understand one another. 'Power' concerns how people and things can align effectively, whereas 'knowledge' concerns how they can align informatively. Some of the same elements and patterns may participate in both kinds of alignment, and an epistemic alignment may itself contribute to an alignment of power or vice versa. Knowledge and power frequently come together in scientific practices, which have not only changed how the world is understood but also influenced people's situations and life possibilities.[42] As a consequence, the critical assessment of scientific claims to (or presumptions of) knowledge must

42. For a more extended and detailed discussion of the intertwining of knowledge and power in scientific practices, see Rouse 1987b: chap. 7.

often be closely intertwined with criticisms of power relations that the associated scientific practices help constitute or sustain. Feminist studies of science have often displayed the most sophisticated grasp of this point, for they have never given up the quest for more adequate and reliable knowledge of the world, while recognizing that claims to knowledge are inevitably entangled in relations of domination and empowerment.

The final two points I want to make about cultural studies are closely connected. Cultural studies take a subversive rather than an antagonistic stance toward some long-standing philosophical questions about science, such as realism and value neutrality; they challenge the formulation of the question rather than proposing an alternative to its traditional answers. This approach is in turn connected to the place of epistemological and political criticism within cultural studies of science. Cultural studies endorse neither the global legitimations of science often put forward by philosophers nor the attempt by many sociologists of science to describe science while bracketing or relativizing any critical assessment of it.

Realism is the view that science (often successfully) aims to provide theories that truthfully represent how the world is independent of human categories, capacities, and interventions. Social constructivists typically reject realism on two counts: the world that science describes is itself socially constituted, and its aims in describing that world are socially specifiable (satisfying interests, sustaining institutions and practices, and so on). Cultural studies of science are better understood as rejecting both realism and the various antirealisms, including social constructivism.[43] Both realists and antirealists propose to explain the content of scientific knowledge, either by its causal connections to real objects or by the social interactions that fix its content. The shared presumption here is that there is a fixed "content" to be explained. Both scientific realists and antirealists presume semantic realism, the thesis that there is an already determinate fact of the matter about what our theories, conceptual schemes, or forms of life "say" about the world. Interpretation must come to an end somewhere, they insist, if not in a world of independently real objects, then in a language, conceptual scheme, social context, or culture.

Cultural studies instead reject the dualism of scheme and content,

43. Rouse (1987b: chap. 5) offers an extended argument rejecting both realism and empiricist or constructivist antirealisms, based on a minimalist or "deflationary" conception of truth. Chapters 7–8 above develop this line of argument further and extend it in new directions.

or context and content, altogether. No determinate scheme or context can fix the content of utterances, and hence it is not possible to get outside of language. How a theory or practice interprets the world is itself inescapably open to further interpretation, with no authority beyond what gets said by whom, when.[44] This position has at least two important consequences in comparison with social constructivism. First, cultural studies can readily speak of statements as true, for 'truth' is a semantic concept that never takes us beyond language. To say that '*p*' is true says no more (but also no less) than saying *p*. Second, it dissolves the boundaries between cultural studies of science and the scientific practices they study. Cultural studies offer interpretations of scientific practices, including the texts and utterances that such practices frequently articulate. But scientific practices are themselves already engaged in such interpretations, in citing, reiterating, criticizing, or extending past practice. As Arthur Fine suggested:

> If science is a performance, then it is one where the audience and crew play as well. Directions for interpretation are also part of the act. If there are questions and conjectures about the meaning of this or that, or its purpose, then there is room for those in the production too. The script, moreover, is never finished, and no past dialogue can fix future action. Such a performance is not susceptible to a reading or interpretation in any global sense, and it picks out its own interpretations, locally, as it goes along.[45]

Cultural studies' interpretive readings are thus part of the culture of science and not an explanation or interpretation of it from "outside." The boundaries between its "inside" and "outside," its centers and its margins, are always themselves at issue in interpretive practice and not something already fixed. The point is not to place all interpretations on a par, for some count as relevant, serious, and significant while others do not. Rather it is to say that which interpretations count in this way, when, and where are themselves part of what is at stake in ongoing interpretation.

What I earlier called the "openness" of scientific practice is crucial here. Internalist history and philosophy of science as well as social constructivism are thus both mistaken when they try to fix what is relevant to the determination of truth, either to reasons and evidence

44. For further discussion of this account of meaning and truth, see Wheeler 1986, 1991; and Rouse 1987b: chap. 5.

45. Fine 1986b: 148.

narrowly construed or to "social factors." One cannot separate the determination of the truth of a scientific claim from the heterogeneous considerations that shape it as a truth claim, one that is intelligible, significant, bearing a (variable) burden of proof, and relevant to various other practices and claims.

Cultural studies likewise try to subvert questions about whether science is (or should be) value-neutral. Traditional discussions of "the" question of value neutrality reify the notion of 'value' just as the realism debates attempt to reify 'truth'.[46] Questions about truth inevitably devolve into multiple questions about significance, relevance, intelligibility, or burden of proof. Similarly, Robert Proctor has argued, the question of value neutrality is not one question but many.[47] Proctor's work thereby opens a significant topic for cultural studies of science, namely, to locate historically and culturally the very conception of scientific research and knowledge as value-free.

The prominence of the term 'value-free' undoubtedly stems from the influence of Max Weber. Ironically, Proctor has shown us, Weber's principal concern was not to keep values from influencing science but the reverse. His advocacy of *Wertfreiheit* was a critique of scientism. But other important concerns have been articulated under this same heading. Against the Nazis' advocacy of a racialized and nationalized science or the Soviet Communist Party's rejection of Mendelian genetics, the notion of value freedom has been timidly invoked to challenge the political censorship of scientific work (timid, for it suggests that if science were not fully and rigidly value-free, it might be appropriately subject to censorship). Similarly timidly, the notion of value freedom has been used to challenge exclusions of scientists on grounds of gender, race, nationality, or political or religious affiliation. A very different use of the conception of "value freedom" has been to draw problematic distinctions of pure from applied, or basic from "mission-oriented" research. Of course, those scientists and their employers whose work is applied or mission-oriented by any plausible criterion have not hesitated to appropriate the legitimating notion of value freedom.

46. Stich (1990) displays a stunning blindness to this parallel. Having powerfully argued that any attempt to fix the intension of 'truth' potentially drives a wedge between " '*p*' is true" and the reasons for believing "*p*," Stich settles on a cultural pluralism about reason by appealing to the values that fix the objects of an individual's or culture's desires, without recognizing that an exactly parallel argument could be developed to fragment the concept of 'value'.

47. Proctor 1991.

Value freedom is also attributed to nature as well as to science. Here we encounter both the modern conception of the "disenchanted" universe, which rejects an ordered cosmos, and the criticisms of vitalism and teleology in biology. This conception of nature as value-free is in direct conflict with the frequent use of scientific work to legitimate or discredit values (for example, the controversies over sociobiology). But what is important for our purposes is that the various conceptions of nature as disenchanted and science as value-free are an important topic for cultural studies, with a rich and contradictory history, and not a framing of its investigations.

These discussions of the concepts of truth and value lead us to the final issue that I take to characterize cultural studies of science. Sociological constructivists frequently insist that they merely describe the ways in which scientific knowledge is socially produced, while bracketing any questions about its epistemic or political worth. In this respect, their work belongs to the tradition that posits value freedom as a scientific ideal. By contrast, cultural studies of science have a stronger reflexive sense of their own cultural and political engagement and typically do not eschew epistemic or political criticism. They find normative issues inevitably at stake in both science and cultural studies of science but see them as arising both locally and reflexively. One cannot not be politically and epistemically engaged.

Two examples of how the burden of proof is determined in AIDS research will illustrate my point and reinforce the earlier claim that cultural studies of science are in the end continuous with the reflexive practice of science itself. Paula Treichler and Cindy Patton have both noted that retrovirologists confidently announced that a sequence of RNA which they had isolated was "the AIDS virus" or "the cause of AIDS" long before anything had been established about its detailed role in the clinical development of the disease or about the presence or absence of cofactors.[48] Within the present scientific climate, the burden of proof falls heavily on the opponents of what Keller has called "master molecule" explanations of biological phenomena; what kind and degree of evidence they and the proponents of such explanations need to provide for their claims differs accordingly among them.[49] Similarly, the widespread scientific discussion of the "African origin" of AIDS has, for historically and politically significant reasons, confronted looser standards of evidence than have other claims about

48. Treichler 1989, 1992; and Patton 1990.
49. Keller 1985: chap. 8.

its epidemiology. Treichler's and Patton's arguments in each case are neither uncritical descriptions of how the scientific burden of proof is assigned nor part of a global relativizing of scientific argument. Instead, they offer a detailed criticism of both how that burden falls and its consequences, via an interpretation of how it was historically constituted. Their argument is not that scientific claims should be rejected for extrascientific reasons but that the local patterns of scientific reasoning and relevance relations need to be reconstructed at specific points.

The critical standpoint afforded by such cultural studies is not that of epistemic sovereignty as inscribed in a "narrative leviathan,"[50] which would legislate for science and culture on the basis of its grasp of the right explanatory factors to account for scientific knowledge without residue. Rather, cultural studies are located within ongoing conflicts over knowledge, power, identity, and possibilities for action. Whatever critical insight and effectiveness they might have must result from their responsiveness to the resonances and tensions among what I have called the alignments and counteralignments shaping an epistemic situation.[51] As described in Chapter 7, epistemic alignments are dynamic and heterogeneous arrays of practices, objects, and communities or solidarities which reinforce, appropriate, or extend one another and thereby constitute knowledge. Cultural studies are reflexive attempts to strengthen, transform, or reconstitute existing alignments or counteralignments by resituating them historically and geographically.

The crucial differences between the normative standpoints of social constructivism and cultural studies of science are succinctly expressed by several of their most prominent practitioners. Pinch sees "the task for the sociologist [being] to try and recapture some of the 'life world' of the scientist—the taken-for-granted practices and interpretations which make available the natural world."[52] The goal of such a "recapture" is to rearrange the relations of authority among disciplines. As Collins and Yearley put it, "SSK wants to use science to weaken natural science in its relationship to social science. . . . We want all cultural endeavors to be seen as equal in their scientific potential."[53] It is instructive to contrast such accounts to Haraway's articulation of a vision for cultural studies:

50. I take this phrase from Traweek 1992.
51. See Chap. 7 above.
52. Pinch 1986: 19.
53. Collins and Yearley 1992b: 383.

> Feminists have to insist on a better account of the world; it is not enough to show radical historical contingency and modes of construction for everything. . . . [So] 'our' problem is how to have *simultaneously* an account of radical historical contingency for all knowledge claims and knowing subjects, a critical practice for recognizing our own 'semiotic technologies' for making meanings, *and* a no-nonsense commitment to faithful accounts of a 'real' world, one that can be partially shared and friendly to earth-wide projects of finite freedom, adequate material abundance, modest meaning in suffering and limited happiness.[54]

To put the difference polemically, social constructivism is antagonistic to the cultural authority claimed by the natural sciences but uncritical of scientific practices. Cultural studies reverse this stance, aiming to participate in constructing reliable and authoritative knowledge of the world by critically engaging with the sciences' practices of making meanings.

54. Haraway 1990: 187.

References

Abir-Am, Pnina. 1985. Themes, Genres, and Orders of Legitimation in the Consolidation of New Scientific Disciplines. *History of Science* 23: 73–117.

Ackermann, Robert. 1985. *Data, Instruments, and Theory*. Princeton: Princeton University Press.

Adas, Michael. 1989. *Machines as the Measure of Men: Science, Technology, and Ideologies of Western Dominance*. Ithaca: Cornell University Press.

Allen, Barry. 1993. *Truth in Philosophy*. Cambridge: Harvard University Press.

Allen, Garland. 1978. *Life Science in the Twentieth Century*. Cambridge: Cambridge University Press.

Ashmore, Malcolm. 1989. *The Reflexive Thesis: Wrighting Sociology of Scientific Knowledge*. Chicago: University of Chicago Press.

Barnes, Barry. 1982. *T. S. Kuhn and Social Science*. London: Macmillan.

Barrett, William, and Paul Roth. 1991. Deconstructing Quarks: Rethinking Sociological Constructions of Science. *Social Studies of Science* 20: 579–632.

Bechtel, William. 1993. Integrating Sciences by Creating New Disciplines: The Case of Cell Biology. *Biology and Philosophy* 8: 277–99.

Berger, Peter, and Thomas Luckmann. 1966. *The Social Construction of Reality*. New York: Doubleday.

Bernal, J. D. 1954. *Science in History*. Cambridge: MIT Press.

———. 1967. *The Social Function of Science*. Cambridge: MIT Press.

Bhaskar, Roy. 1986. *Scientific Realism and Human Emancipation*. London: Verso.

Biagioli, Mario. 1993. *Galileo, Courtier*. Chicago: University of Chicago Press.

Birke, Lynda. 1986. *Women, Feminism, and Biology*. Brighton: Wheatsheaf.

Bleier, Ruth. 1984. *Science and Gender: A Critique of Biology and Its Theories on Women*. New York: Pergamon.

Bloor, David. 1983. *Wittgenstein: A Social Theory of Knowledge*. New York: Columbia University Press.

———. 1991. *Knowledge and Social Imagery*. 2d ed. Chicago: University of Chicago Press.

———. 1992. Left and Right Wittgensteinians. In Pickering 1992: 266–82.

Boyd, Richard. 1984. The Current Status of Scientific Realism. In Leplin 1984: 41–82.

———. 1988. How to Be a Moral Realist. In *Essays on Moral Realism,* ed. Geoffrey Sayre-McCord, 181–228. Ithaca: Cornell University Press.

Brandom, Robert. 1976. Truth and Assertibility. *Journal of Philosophy* 73: 137–89.

———. 1979. Freedom and Constraint by Norms. *American Philosophical Quarterly* 16: 187–96.

———. 1994. *Making It Explicit: Reasoning, Representing, and Discursive Commitment.* Cambridge: Harvard University Press.

Brown, James Robert, ed. 1984. *Scientific Rationality: The Sociological Turn.* Dordrecht: D. Reidel.

Butler, Judith. 1987. Variations on Sex and Gender: Beauvoir, Wittig, and Foucault. In *Feminism as Critique,* ed. Seyla Benhabib and Drucilla Cornell, 128–42. Minneapolis: University of Minnesota Press.

———. 1990. *Gender Trouble.* New York: Routledge.

———. 1993. *Bodies That Matter.* New York: Routledge.

Carnap, Rudolf. 1928. *The Logical Structure of the World.* Trans. R. George. Berkeley: University of California Press.

Carr, David. 1986. *Time, Narrative, and History.* Bloomington: Indiana University Press.

Cartwright, Nancy. 1983. *How the Laws of Physics Lie.* Oxford: Clarendon.

———. 1989. *Nature's Capacities and Their Measurement.* Oxford: Oxford University Press.

———. 1994. Fundamentalism vs. the Patchwork of Laws. *Proceedings of the Aristotelian Society* 94: 279–92.

Cat, Jordi, Nancy Cartwright, and Hasok Chang. Forthcoming. Otto Neurath: Politics and the Unity of Science. In Galison and Stump Forthcoming.

Churchland, Paul. 1979. *Scientific Realism and the Plasticity of Mind.* Cambridge: Cambridge University Press.

———. 1989. *A Neurocomputational Perspective: The Nature of Mind and the Structure of Science.* Cambridge: MIT Press.

———. 1992. A Deeper Unity: Some Feyerabendian Themes in Neurocomputational Form. In *Cognitive Models of Science,* ed. Ronald Giere, 341–66. Minneapolis: University of Minnesota Press.

Churchland, Paul, and Clifford Hooker, ed. 1985. *Images of Science: Essays on Realism and Empiricism.* Chicago: University of Chicago Press.

Clifford, James, and George Marcus, ed. 1986. *Writing Culture: The Poetics and Politics of Ethnography.* Berkeley: University of California Press.

Code, Lorraine. 1991. *What Can She Know?* Ithaca: Cornell University Press.

Coffa, J. Alberto. 1991. *To the Vienna Station: The Semantic Tradition from Kant to Carnap,* ed. Linda Wessels. Cambridge: Cambridge University Press.

Collins, Harry. 1983. An Empirical Relativist Programme in the Sociology of Scientific Knowledge. In Knorr-Cetina and Mulkay 1983: 85–113.

———. 1992. *Changing Order: Replication and Induction in Scientific Practice.* 2d ed. Chicago: University of Chicago Press.

Collins, Harry, and Steven Yearley. 1992a. Epistemological Chicken. In Pickering 1992: 301–26.

———. 1992b. Journey into Space. In Pickering 1992: 369–89.

Crosby, Alfred. 1986. *Ecological Imperialism: The Biological Expansion of Europe, 900–1900.* Cambridge: Cambridge University Press.
Davidson, Donald. 1984. *Inquiries into Truth and Interpretation.* Oxford: Oxford University Press.
———. 1986. A Nice Derangement of Epitaphs. In *Truth and Interpretation: Essays on the Philosophy of Donald Davidson,* ed. Ernest Lepore, 433–46. Oxford: Basil Blackwell.
Dennett, Daniel. 1987. *The Intentional Stance.* Cambridge: MIT Press.
———. 1991a. *Consciousness Explained.* Boston: Little, Brown.
———. 1991b. Real Patterns. *Journal of Philosophy* 91: 27–51.
Derrida, Jacques. 1973. *Speech and Phenomena.* Trans. David Allison. Evanston: Northwestern University Press.
———. 1976. *Of Grammatology.* Trans. Gayatri Spivak. Chicago: University of Chicago Press.
———. 1988. *Limited Inc.* Trans. Samuel Weber and Jeffrey Mehlman. Evanston: Northwestern University Press.
Dretske, Frederick. 1981. *Knowledge and the Flow of Information.* Cambridge: MIT Press.
Dreyfus, Hubert. 1980. Holism and Hermeneutics. *Review of Metaphysics* 34: 3–23.
———. 1984. Why Current Studies of Human Capacities Can Never Be Made Scientific. *Berkeley Cognitive Science Report* 11: 1–17.
———. 1991. *Being-in-the-World: A Commentary on Heidegger's "Being and Time," Division I.* Cambridge: MIT Press.
———. 1992. The Hermeneutic Approach to Intentionality. Paper presented to 18th Annual Meeting of the Society for Philosophy and Psychology, Montreal, Quebec.
Dupre, John. 1993. *The Disorder of Things: Metaphysical Foundations of the Disunity of Science.* Cambridge: Harvard University Press.
During, Simon, ed. 1993. *The Cultural Studies Reader.* New York: Routledge.
Fausto-Sterling, Anne. 1985. *Myths of Gender.* New York: Basic.
Feyerabend, Paul. 1975. *Against Method.* London: NLB.
———. 1987. *Farewell to Reason.* London: Verso.
Fine, Arthur. 1986a. The Natural Ontological Attitude. In Fine 1986c: 112–35.
———. 1986b. And Not Anti-Realism Either. In Fine 1986c: 136–50.
———. 1986c. *The Shaky Game: Einstein, Realism, and the Quantum Theory.* Chicago: University of Chicago Press.
———. 1986d. Unnatural Attitudes: Realist and Instrumentalist Attachments to Science. *Mind* 95: 149–79.
———. 1991. Piecemeal Realism. *Philosophical Studies* 61: 79–96.
———. Forthcoming. Science Made Up: Constructivist Sociology of Scientific Knowledge. In Galison and Stump forthcoming.
Fleck, Ludwik. 1979. *Genesis and Development of a Scientific Fact.* Trans. Fred Bradley and Thaddeus Trenn. Chicago: University of Chicago Press.
Fodor, Jerry. 1980. Methodological Solipsism Considered as a Research Strategy in Cognitive Psychology. *Behavioral and Brain Sciences* 3: 417–24.
Foucault, Michel. 1977. *Discipline and Punish.* Trans. Alan Sheridan. New York: Random House.
———. 1978. *The History of Sexuality,* vol. 1. Trans. Robert Hurley. New York: Random House.

———. 1980. *Power/Knowledge*. New York: Random House.

Franklin, Allen. 1987. *The Neglect of Experiment*. Cambridge: Cambridge University Press.

Fraser, Nancy. 1988. Solidarity or Singularity? Richard Rorty between Romanticism and Technocracy. *Praxis International* 8: 257–72.

———. 1989. *Unruly Practices*. Minneapolis: University of Minnesota Press.

Fraser, Nancy, and Linda Nicholson. 1988. Social Criticism without Philosophy: An Encounter between Feminism and Postmodernism. In *Universal Abandon? The Politics of Postmodernism*, ed. Andrew Ross, 83–104. Minneapolis: University of Minnesota Press.

Frege, Gottlob. 1970. On Sense and Reference. Trans. Max Black. In *Philosophical Writings of Gottlob Frege*, ed. Peter Geach and Max Black, 56–78. Oxford: Basil Blackwell.

Friedman, Michael. 1984. Critical notice of *Philosophical Papers, Moritz Schlick. Philosophy of Science* 50: 498–514.

———. 1987. Carnap's *Aufbau* Reconsidered. *Nous* 21: 521–45.

Friedman, Robert Marc. 1989. *Appropriating the Weather: Vilhelm Bjerknes and the Construction of a Modern Meteorology*. Ithaca: Cornell University Press.

Fuller, Steve. 1988. *Social Epistemology*. Bloomington: Indiana University Press.

———. 1989. *Philosophy of Science and Its Discontents*. Boulder, Colo.: Westview.

———. 1992a. Being There with Thomas Kuhn. *History and Theory* 31: 241–75.

———. 1992b. Epistemology Radically Naturalized: Recovering the Normative, the Experimental, and the Social. In *Cognitive Models of Science*, ed. Ronald N. Giere, 427–59. Minneapolis: University of Minnesota Press.

———. 1992c. Social Epistemology and the Research Agenda of Science Studies. In Pickering 1992: 390–428.

Galison, Peter. 1985. Bubble Chambers and the Experimental Workplace. In *Observation, Experiment, and Hypothesis in Modern Physical Science*, ed. Peter Achinstein and Owen Hannaway, 309–73. Cambridge: MIT Press.

———. 1987. *How Experiments End*. Chicago: University of Chicago Press.

———. 1988. History, Philosophy, and the Central Metaphor. *Science in Context* 2: 200–201.

———. 1990. Aufbau/Bauhaus: Logical Positivism and Architectural Modernism. *Critical Inquiry* 16: 709–52.

———. Forthcoming. *Image and Logic: The Material Culture of Modern Physics*.

Galison, Peter, and Bruce Hevly. 1992. *Big Science: The Growth of Large-Scale Research*. Stanford: Stanford University Press.

Galison, Peter, and David Stump. Forthcoming. *The Disunity of Science: Boundaries, Contexts, and Power*. Stanford: Stanford University Press.

Ginzberg, Ruth. 1987. Uncovering Gynocentric Science. *Hypatia* 2: 89–106.

Gooding, David, Trevor Pinch, and Simon Schaffer. 1989. *The Uses of Experiment*. Cambridge: Cambridge University Press.

Goody, Jack. 1977. *The Domestication of the Savage Mind*. Cambridge: Cambridge University Press.

Griffin, Susan. 1978. *Woman and Nature: The Roaring inside Her*. New York: Harper & Row.

Grossberg, Lawrence, Cary Nelson, and Paula Treichler. 1992. *Cultural Studies*. New York: Routledge.

Gutting, Gary. 1980. *Paradigms and Revolutions: Applications and Appraisals of*

Thomas Kuhn's Philosophy of Science. Notre Dame: University of Notre Dame Press.

Habermas, Jürgen. 1982. Modernity versus Postmodernity. *New German Critique* 26: 13–30.

———. 1986. Taking Aim at the Heart of the Present. In Hoy 1986: 103–8.

Hacking, Ian. 1982. Language, Truth, and Reason. In *Rationality and Relativism*, ed. Martin Hollis and Steven Lukes, 48–66. Cambridge: MIT Press.

———. 1983. *Representing and Intervening*. Cambridge: Cambridge University Press.

———. 1988. The Participant Irrealist at Large in the Laboratory. *British Journal for the Philosophy of Science* 39: 277–94.

———. 1992. The Self-Vindication of the Laboratory Sciences. In Pickering 1992: 29–64.

Haraway, Donna. 1981–82. The High Cost of Information in Post-World War II Evolutionary Biology: Ergonomics, Semiotics, and the Sociobiology of Communication Systems. *Philosophical Forum* 13: 244–78.

———. 1989. *Primate Visions: Gender, Race, and Nature in the World of Modern Science*. New York: Routledge.

———. 1990. *Simians, Cyborgs, and Women*, New York: Routledge.

———. 1992a. Otherworldly Conversations; Terran Topics; Local Terms. *Science as Culture* 3: 64–98.

———. 1992b. The Promises of Monsters: A Regenerative Politics for Inappropriate/d Others. In Grossberg, Nelson, and Treichler 1992: 295–337.

———. 1994. A Game of Cat's Cradle: Science Studies, Feminist Theory, Cultural Studies. *Configurations* 2: 59–71.

———. Forthcoming. Modest Witness @ Second Millenium: FemaleMan-c Meets OncoMouse.™

Harding, Sandra. 1986. *The Science Question in Feminism*. Ithaca: Cornell University Press.

———. 1991. *Whose Science? Whose Knowledge?* Ithaca: Cornell University Press.

Hardwig, John. 1991. The Role of Trust in Knowledge. *Journal of Philosophy* 88: 693–708.

Harwood, Jonathan. 1993. *Styles of Scientific Thought: The German Genetics Community, 1900–1933*. Chicago: University of Chicago Press.

Heidegger, Martin. 1962. *Being and Time*. Trans. John Macquarrie and Edward Robinson. New York: Harper & Row.

———. 1977. *Basic Writings*, ed. David Farrell Krell. New York: Harper & Row.

Heldke, Lisa. 1988. Recipes for Theory Making. *Hypatia* 3: 15–30.

Hesse, Mary. 1980. *Revolutions and Reconstructions in the Philosophy of Science*. Bloomington: Indiana University Press.

———. 1986. Changing Concepts and Social Order. *Social Studies of Science* 16: 714–26.

Horwich, Paul. 1990. *Truth*. Oxford: Basil Blackwell.

Hoy, David. 1986. *Foucault: A Critical Reader*. Oxford: Basil Blackwell.

Hrdy, Sarah Blaffer. 1981. *The Woman That Never Evolved*. Cambridge: Harvard University Press.

Hubbard, Ruth. 1990. *The Politics of Women's Biology*. New Brunswick: Rutgers University Press.

Hume, David. 1888. *A Treatise of Human Nature*. Oxford: Clarendon.

Husserl, Edmund. 1970. *Logical Investigations.* 2 vols. Trans. J. N. Findlay. London: Routledge & Kegan Paul.

Irigaray, Luce. 1987. Le sujet de la science est-il sexué?/Is the Subject of Science Sexed? Trans. Carol M. Bove. *Hypatia* 2: 65–87.

Jacob, François. 1981. *The Possible and the Actual.* Seattle: University of Washington Press.

Jameson, Fredric. 1984. Postmodernism, or the Cultural Logic of Late Capitalism. *New Left Review* 146: 53–92.

Kant, Immanuel. 1965. *Critique of Pure Reason.* Trans. Norman Kemp Smith. New York: St. Martin's.

Kay, Lily. 1993. *The Molecular Vision of Life: Caltech, the Rockefeller Foundation, and the Rise of the New Biology.* Oxford: Oxford University Press.

Keller, Evelyn Fox, 1983. *A Feeling for the Organism: The Life and Work of Barbara McClintock.* San Francisco: W. H. Freeman.

———. 1985. *Reflections on Gender and Science.* New Haven: Yale University Press.

———. 1990. Physics and the Emergence of Molecular Biology. *Journal of the History of Biology* 23: 389–409.

———. 1992. *Secrets of Life, Secrets of Death: Essays on Language, Gender, and Science.* New York: Routledge.

Kohler, Robert. 1991. *Partners in Science: Foundations and Natural Scientists, 1900–1945.* Chicago: University of Chicago Press.

———. 1994. *Lords of the Fly: Drosophila Genetics and the Experimental Life.* Chicago: University of Chicago Press.

Knorr-Cetina, Karin. 1981. *The Manufacture of Knowledge: An Essay on the Constructive and Contextual Nature of Science.* Oxford: Pergamon.

Knorr-Cetina, Karin, and Michael Mulkay. 1983. *Science Observed: Perspectives on the Social Study of Science.* London: Sage.

Kripke, Saul. 1982. *Wittgenstein on Rules and Private Language.* Cambridge: Harvard University Press.

Kuhn, Thomas. 1970. *The Structure of Scientific Revolutions.* 2d ed. Chicago: University of Chicago Press.

———. 1977. Objectivity, Value Judgment, and Theory Choice. In *The Essential Tension,* 320–39. Chicago: University of Chicago Press, 1977.

Lakatos, Imre. 1978. *The Methodology of Scientific Research Programmes,* ed. John Worrall and Gregory Currie. Cambridge: Cambridge University Press.

Latour, Bruno. 1983. Give Me a Laboratory and I Will Raise the World. In Knorr-Cetina and Mulkay 1983: 141–70.

———. 1984. *Les microbes: Guerre et paix, suivi par irreductions.* Paris: Metailie. *The Pasteurization of France.* Trans. Alan Sheridan and John Law. Cambridge: Harvard University Press, 1988.

———. 1987. *Science in Action.* Cambridge: Harvard University Press.

———. 1988. The Politics of Explanation: An Alternative. In *Knowledge and Reflexivity,* ed. Steve Woolgar, 155–76. Beverley Hills: Sage.

———. 1990. Postmodern? No, Amodern: Steps toward an Anthropology of Science. *Studies in History and Philosophy of Science* 21: 145–71.

———. 1992. One More Turn after the Social Turn. In McMullin 1992: 272–94.

———. 1993. *We Have Never Been Modern.* Trans. Catherine Porter. Cambridge: Harvard University Press.

Latour, Bruno, and Steve Woolgar. 1986. *Laboratory Life: The Construction of Scientific Facts,* 2d ed. Princeton: Princeton University Press.

Laudan, Larry. 1977. *Progress and Its Problems.* Berkeley: University of California Press.

———. 1981. A Confutation of Convergent Realism. *Philosophy of Science* 48: 19–49.

———. 1984. *Science and Values.* Berkeley: University of California Press.

———. 1987. Progress or Rationality? The Prospects for Normative Naturalism. *American Philosophical Quarterly* 24: 19–31.

Laudan, Larry, Arthur Donovan, Rachel Laudan, Peter Barker, Harold Brown, Jerrold Leplin, Paul Thagard, and Steve Wykstra. 1986. Scientific Change: Philosophical Models and Historical Research. *Synthese* 69: 141–223.

Lenoir, Timothy. 1994. Was the Last Turn the Right Turn? The Semiotic Turn and A. J. Greimas. *Configurations* 2: 119–36.

Leplin, Jarrett. 1984. *Scientific Realism.* Berkeley: University of California Press.

Longino, Helen. 1990. *Science as Social Knowledge.* Princeton: Princeton University Press.

Lynch, Michael. 1985. *Art and Artefact in a Laboratory Science.* London: Routledge.

———. 1992. Extending Wittgenstein: The Pivotal Move from Epistemology to Sociology of Science. In Pickering 1992: 215–65.

Lyotard, Jean-François. 1984. *The Postmodern Condition: A Report on Knowledge.* Trans. Geoff Bennington and Brian Massumi. Minneapolis: University of Minnesota Press.

MacIntyre, Alasdair. 1980. Epistemological Crises, Dramatic Narrative, and the Philosophy of Science. In Gutting 1980: 54–74.

———. 1981. *After Virtue.* Notre Dame: University of Notre Dame Press.

MacKenzie, Donald, and Barry Barnes. 1979. Scientific Judgment: The Biometry-Mendelism Controversy. In *Natural Order: Historical Studies of Scientific Culture,* ed. Barry Barnes and Steven Shapin. Beverly Hills: Sage.

Mannheim, Karl. 1952. *Essays on the Sociology of Knowledge.* London: Routledge & Kegan Paul.

Marcus, George, and Michael Fischer. 1986. *Anthropology as Cultural Critique.* Chicago: University of Chicago Press.

McMullin, Ernan. 1992. *The Social Dimensions of Science.* Notre Dame: University of Notre Dame Press.

McNeill, William. 1976. *Plagues and Peoples.* Garden City, N.Y.: Anchor/Doubleday.

Merleau-Ponty, Maurice. 1962. *Phenomenology of Perception.* Trans. Colin Smith. London: Routledge & Kegan Paul.

Merton, Robert. 1973. *The Sociology of Science: Theoretical and Empirical Investigations.* Chicago: University of Chicago Press.

Miller, Richard. 1987. *Fact and Method.* Princeton: Princeton University Press.

Millikan, Ruth. 1984. *Language, Thought, and Other Biological Categories.* Cambridge: MIT Press.

Mink, Louis O. 1987. *Historical Understanding.* Ithaca: Cornell University Press.

Mitchell, Sam. 1988. Constructive Empiricism and Antirealism. In *PSA 1988,* vol. 1, ed. Arthur Fine and Jarrett Leplin, 174–80. East Lansing, Mich.: Philosophy of Science Association.

Musgrave, Alan. 1980. Kuhn's Second Thoughts. In Gutting 1980: 39–53.

Nelson, Lynn Hankinson. 1990. *Who Knows? From Quine to a Feminist Empiricism.* Philadelphia: Temple University Press.

Okrent, Mark. 1988. *Heidegger's Pragmatism.* Ithaca: Cornell University Press.

———. 1989. The Metaphilosophical Consequences of Pragmatism. In *The Institution of Philosophy,* ed. Avner Cohen and Marcel Dascal, 177–98. Totowa, N.J.: Rowman and Allanheld.

———. 1991. Teleological Underdetermination. *American Philosophical Quarterly* 28: 147–56.

———. 1993. The Truth, the Whole Truth, and Nothing but the Truth. *Inquiry* 36: 381–404.

———. Forthcoming. *Teleology, Language, and the Mental.*

Olby, Robert. 1979. Mendel No Mendelian? *History of Science* 17: 53–72.

———. 1985. *Origins of Mendelism.* 2d ed. Chicago: University of Chicago Press.

Patton, Cindy. 1990. *Inventing AIDS.* New York: Routledge.

Pickering, Andrew. 1984. *Constructing Quarks: A Sociological History of Particle Physics.* Chicago: University of Chicago Press.

———. 1989a. Big Science as a Form of Life. In *The Restructuring of the Physical Sciences in Europe and the United States, 1945–1960,* ed. Michelangelo de Maria, Mario Grilli, and Fabio Sebastiani, Singapore: World Scientific Publishing.

———. 1989b. Living in the Material World. In Gooding 1989: 275–97.

———. 1992. *Science as Practice and Culture.* Chicago: University of Chicago Press.

———. 1995. *The Mangle of Practice.* Chicago: University of Chicago Press.

Pinch, Trevor. 1986. *Confronting Nature.* Dordrecht: D. Reidel.

Polanyi, Michael. 1958. *Personal Knowledge.* Chicago: University of Chicago Press.

Popper, Karl. 1962. *Conjectures and Refutations.* New York: Basic.

Proctor, Robert. 1991. *Value-Free Science? Purity and Power in Modern Knowledge.* Cambridge: Harvard University Press.

Putnam, Hilary. 1978. *Meaning and the Moral Sciences.* London: Routledge & Kegan Paul.

Quine, Willard van Orman. 1953. Two Dogmas of Empiricism. In *From a Logical Point of View,* 20–46. Cambridge: Harvard University Press.

Rabinow, Paul. 1992. Artificiality and Enlightenment: From Sociobiology to Biosociality. In *Incorporations,* ed. Jonathan Crary and Sanford Kwinter, 234–53. Cambridge: MIT Press.

Ramberg, Bjorn. 1989. *Donald Davidson's Philosophy of Language.* Oxford: Basil Blackwell.

Rheinberger, Hans-Jörg. 1992a. Experiment, Difference, and Writing: I. Tracing Protein Synthesis. *Studies in History and Philosophy of Science* 23: 305–31.

———. 1992b. Experiment, Difference, and Writing: II. The Laboratory Production of Transfer RNA, *Studies in History and Philosophy of Science* 23: 389–422.

———. 1992c. *Experiment, Differenz, Schrift.* Marburg: Basiliskenpresse.

———. 1994. Experimental Systems: Historiality, Narration and Deconstruction. In *Science in Context* 7: 65–81.

Richardson, Alan. 1990. How Not to Russell Carnap's *Aufbau.* In *PSA 1990,* vol. 1, ed. Arthur Fine, Mickey Forbes, and Linda Wessels, 3–14. East Lansing, Mich.: Philosophy of Science Association.

Rorty, Richard. 1979. *Philosophy and the Mirror of Nature.* Princeton: Princeton University Press.

———. 1982. *Consequences of Pragmatism.* Minneapolis: University of Minnesota Press.

———. 1989. *Contingency, Irony, and Solidarity*. Cambridge: Cambridge University Press.

———. 1991. *Objectivity, Relativism, and Truth: Philosophical Papers*, vol. 1. Cambridge: Cambridge University Press.

Rosaldo, Renato. 1989. *Culture and Truth*. Boston: Beacon.

Roth, Paul. 1987. *Meaning and Method in the Social Sciences*. Ithaca: Cornell University Press.

Rouse, Joseph. 1987a. Husserlian Phenomenology and Scientific Realism. *Philosophy of Science* 54: 222–32.

———. 1987b. *Knowledge and Power: Toward a Political Philosophy of Science*. Ithaca: Cornell University Press.

———. 1991a. Indeterminacy, Empirical Evidence, and Methodological Pluralism. *Synthese* 86: 443–65.

———. 1991b. Interpretation in Natural and Human Science. In *The Interpretive Turn*, ed. David Hiley, James Bohman, and Richard Shusterman, 42–56. Ithaca: Cornell University Press.

———. 1991c. Policing Knowledge: Disembodied Policy for Embodied Knowledge. *Inquiry* 34: 353–64.

———. 1993. Foucault and the Natural Sciences. In *Foucault and the Critique of Institutions*, ed. John Caputo and Mark Yount, 137–62. State College: Pennsylvania State University Press.

———. Forthcoming a. Beyond Epistemic Sovereignty. In Galison and Stump forthcoming.

———. Forthcoming b. Feminism and the Social Construction of Scientific Knowledge. In *A Dialogue Concerning Feminism, Science, and the Philosophy of Science*, ed. Lynn Hankinson Nelson and Jack Nelson. Dordrecht: Kluwer.

Ryckman, Thomas. 1991a. *Condition Sine Qua Non? Zuordnung* in the Early Epistemology of Cassirer and Schlick. *Synthese* 88: 57–95.

———. 1991b. Designation and Convention: A Chapter of Early Logical Empiricism. In *PSA 1990*, vol. 2, ed. Arthur Fine, Mickey Forbes, and Linda Wessels, 149–57. East Lansing, Mich.: Philosophy of Science Association.

Sapp, Jan. 1987. *Beyond the Gene: Cytoplasmic Inheritance and the Struggle for Authority in Genetics*. Oxford: Oxford University Press.

Searle, John. 1983. *Intentionality: An Essay in the Philosophy of Mind*. Cambridge: Cambridge University Press.

Scheffler, Israel. 1982. *Science and Subjectivity*, 2d ed. Indianapolis: Hackett.

Sellars, Wilfred. 1963. *Science, Perception, and Reality*. London: Routledge & Kegan Paul.

———. 1968. *Science and Metaphysics*. London: Routledge & Kegan Paul.

Shapere, Dudley. 1984. *Reason and the Search for Knowledge*. Dordrecht: D. Reidel.

———. 1986. External and Internal Factors in the Development of Science. *Science and Technology Studies* 4: 1–9, 19–23.

Shapin, Steven. 1982. History of Science and Its Sociological Reconstructions. *History of Science* 20: 157–211.

———. 1994. *A Social History of Truth: Civility and Science in Seventeenth-Century Science*. Chicago: University of Chicago Press.

Shapin, Steven, and Simon Schaffer. 1985. *Leviathan and the Air-Pump*. Princeton: Princeton University Press.

Smith, Crosbie, and Norton Wise. 1989. *Energy and Empire: A Biographical Study of Lord Kelvin*. Cambridge: Cambridge University Press.

Smocovitis, Vassiliki Betty. 1992. Unifying Biology: The Evolutionary Synthesis and Evolutionary Biology. *Journal of the History of Biology* 26: 1–65.

———. Forthcoming. *Unifying Biology: The Evolutionary Synthesis and Evolutionary Biology*. Princeton: Princeton University Press.

Sneed, Joseph. 1971. *The Logical Structure of Mathematical Physics*. Dordrecht: D. Reidel.

Stich, Stephen. 1990. *The Fragmentation of Reason*. Cambridge: MIT Press.

Stump, David. 1992. Naturalized Philosophy of Science with a Plurality of Methods. *Philosophy of Science* 59: 456–60.

Tarski, Alfred. 1944. The Semantic Conception of Truth. *Philosophy and Phenomenological Research* 4: 341–75.

Taylor, Charles. 1986. Foucault on Freedom and Truth. In Hoy 1986: 69–102.

———. 1991. The Dialogical Self. In *The Interpretive Turn*, ed. David Hiley, James Bohman, and Richard Shusterman, 304–14. Ithaca: Cornell University Press.

Todes, Samuel. 1966. A Comparative Phenomenology of Perception and Imagination, Part I. *Journal of Existentialism* 6: 253–68.

Traweek, Sharon. 1988. *Beamtimes and Lifetimes*. Cambridge: Harvard University Press.

———. 1992. Border Crossings: Narrative Strategies in Science Studies and among Physicists in Tsukuba Science City, Japan. In Pickering 1992: 429–65.

Treichler, Paula. 1988. AIDS, Homophobia, and Biomedical Discourse: An Epidemic of Signification. In *AIDS: Cultural Analysis/Cultural Activism*, ed. Douglas Crimp, 31–70. Cambridge: MIT Press.

———. 1989. AIDS and HIV Infection in the Third World: First World Chronicles. In *Remaking History*, ed. Barbara Kruger and Phil Mariani, 31–86. New York: Dia Art Foundation.

———. 1991. How to Have Theory in an Epidemic: The Evolution of AIDS Treatment Activism. In *Technoculture*, ed. Constance Penley and Andrew Ross, 57–106. Minneapolis: University of Minnesota Press.

———. 1992. AIDS, HIV, and the Cultural Construction of Reality. In *The Time of AIDS*, ed. Gilbert Herdt and Shirley Lindenbaum, 65–98. Beverly Hills: Sage.

Tuana, Nancy. 1988. Feminism and Science II. *Hypatia* 3: 1–168.

Turner, Stephen. 1994. *The Social Theory of Practices*. Chicago: University of Chicago Press.

van Fraassen, Bas. 1980. *The Scientific Image*. Oxford: Clarendon.

Wartenburg, Thomas. 1990. *The Forms of Power*. Philadelphia: Temple University Press.

Werskey, Gary. 1978. *The Visible College*. New York: Holt, Rinehart, and Winston.

Wheeler, Samuel C., III. 1986. Indeterminacy of French Interpretation: Derrida and Davidson. In *Truth and Interpretation: Perspectives on the Philosophy of Donald Davidson*, ed. Ernest LePore, 477–94. Oxford: Blackwell.

———. 1989. Metaphor according to Davidson and de Man. In *Redrawing the Lines: Analytic Philosophy, Deconstruction, and Literary Theory*, ed. Reed Dasenbrock, 116–39. Minneapolis: University of Minnesota Press.

———. 1991. True Figures: Metaphor, Social Relations, and the Sorites. In *The Interpretive Turn*, ed. David Hiley, James Bohman, and Richard Shusterman, 197–217. Ithaca: Cornell University Press.

Williams, Michael. 1986. Do We (Epistemologists) Need a Theory of Truth? *Philosophical Topics* 14: 223–42.

———. 1991. *Unnatural Doubts: Epistemological Realism and the Basis of Scepticism.* Oxford: Basil Blackwell.

Winch, Peter. 1958. *The Idea of a Social Science and Its Relation to Philosophy.* London: Routledge & Kegan Paul.

———. 1964. Understanding a Primitive Society. *American Philosophical Quarterly* 1: 307–24.

Winner, Langdon. 1977. *Autonomous Technology.* Cambridge: MIT Press.

Wittgenstein, Ludwig. 1953. *Philosophical Investigations.* Oxford: Basil Blackwell.

Woolgar, Steve. 1983. Irony in the Social Study of Science. In Knorr-Cetina and Mulkay 1983: 239–66.

———. 1988a. *Knowledge and Reflexivity: New Frontiers in the Sociology of Knowledge.* Beverly Hills: Sage.

———. 1988b. Reflexivity Is the Ethnographer of the Text. In Woolgar 1988a: 15–34.

———. 1988c. *Science: The Very Idea.* London: Tavistock.

———. 1991. The Very Idea of Social Epistemology. *Inquiry* 34: 377–89.

———. 1992. Some Remarks about Positionism: A Reply to Collins and Yearley. In Pickering 1992: 327–42.

Wylie, Alison. 1991. Gender Theory and the Archaeological Record: Why Is There No Archaeology of Gender? In *Engendering Archaeology: Women and Prehistory,* ed. Joan M. Gero and Margaret W. Conkey, 31–54. Oxford: Blackwell.

———. 1992. The Interplay of Evidential Constraints and Political Interests: Recent Archaeological Research on Gender. *American Antiquity* 57: 15–35.

———. Forthcoming. The Constitution of Archaeological Evidence: Gender Politics and Science. In Galison and Stump forthcoming.

Index